T0207142

Unlocking dbt

Design and Deploy Transformations in Your Cloud Data Warehouse

Cameron Cyr
Dustin Dorsey

Apress®

Unlocking dbt: Design and Deploy Transformations in Your Cloud Data Warehouse

Cameron Cyr
Chapel Hill, TN, USA

Dustin Dorsey
Murfreesboro, TN, USA

ISBN-13 (pbk): 978-1-4842-9699-8
ISBN-13 (electronic): 978-1-4842-9703-2
https://doi.org/10.1007/978-1-4842-9703-2

Managing Director, Apress Media LLC: Welmoed Spahr
Acquisitions Editor: Jonathan Gennick
Development Editor: Laura Berendson
Editorial Project Manager: Shaul Elson

Cover image by fabrlkaslmf on Freepik

Distributed to the book trade worldwide by Springer Science+Business Media LLC, 1 New York Plaza, Suite 4600, New York, NY 10004. Phone 1-800-SPRINGER, fax (201) 348-4505, e-mail orders-ny@springer-sbm.com, or visit www.springeronline.com. Apress Media, LLC is a California LLC and the sole member (owner) is Springer Science + Business Media Finance Inc (SSBM Finance Inc). SSBM Finance Inc is a **Delaware** corporation.

For information on translations, please e-mail booktranslations@springernature.com; for reprint, paperback, or audio rights, please e-mail bookpermissions@springernature.com.

Apress titles may be purchased in bulk for academic, corporate, or promotional use. eBook versions and licenses are also available for most titles. For more information, reference our Print and eBook Bulk Sales web page at http://www.apress.com/bulk-sales.

Any source code or other supplementary material referenced by the author in this book is available to readers on GitHub (https://github.com/Apress). For more detailed information, please visit https://www.apress.com/gp/services/source-code.

Paper in this product is recyclable

To my wife Tristyn, your belief in me has fueled my ambitions and pushed me to achieve more than I ever thought possible. I have found strength with you by my side, and I have discovered a love that knows no bounds.

—Cameron

To my wife Sarah and my wonderful children Zoey, Bennett, and Ellis, the unwavering pillars of my life. I dedicate this to you all for your endless patience, encouragement, and belief in me as I went through this journey.

—Dustin

Table of Contents

About the Authors

Cameron Cyr is a data enthusiast who has spent his career developing data systems enabling valuable use cases such as analytics and machine learning. During his career, he has placed a focus on building reliable and scalable data systems with an emphasis on data quality. He is active in the data community and is one of the organizers and founders of the Nashville Data Engineering Group.

Dustin Dorsey is a data leader and architect who has been building and managing data solutions for over 15 years. He has worked with several startups and Fortune 500 companies in the healthcare, manufacturing, and biotech spaces designing and building out data infrastructure and advanced analytics. Dustin is an international speaker, mentor, and well-respected leader in the data community. He has previously organized several data community events and user groups and currently is one of the founders and organizers of the Nashville Data Engineering Group. Dustin is also one of the authors of the popular Apress book *Pro Database Migration to Azure*.

About the Technical Reviewer

Alice Leach is a Data Engineer at Whatnot Inc., a social marketplace that enables collectors and enthusiasts to connect, buy, and sell verified products. She began her career in science, obtaining an MChem in Chemistry from the University of Oxford (2012) and a PhD in Interdisciplinary Materials Science from Vanderbilt University (2017). She then spent four years as a Research Assistant Professor at the Vanderbilt Institute of Nanoscale Science and Engineering. She transitioned from academia to data in 2021, working first as a data scientist and then a data engineer. Her current work at Whatnot focuses on designing and building robust, self-service data workflows using a modern data stack.

Acknowledgments

The opportunity to write this book is not something I anticipated, but I have much to be thankful for because of and about this opportunity. To start, I want to thank my wife, Tristyn, who always pushes me to grow personally and professionally. More than anything, I want to thank her for supporting me as I spent countless nights and weekends locked away in my office writing. Tristyn, I thank you for your unwavering love, for being my confidante, my best friend, and my beloved partner. I love you more than anything.

Of course, I also want to thank my coauthor, Dustin Dorsey. From the first time we met, Dustin has continuously challenged me to grow in ways that I never thought were achievable. One of those challenges that Dustin presented to me is this book, and I am eternally grateful that together we were able to seize the opportunity to publish one of the first books about dbt. As a colleague, Dustin is the most brilliant data architect that I have ever had the pleasure to work with. As a friend, I thank him for encouraging me during the times that writing became a struggle.

To conclude, I want to also thank the many colleagues that I have worked with throughout my career or within the data community. Without the individual contributions that you all have made to inspire, encourage, and motivate me, I wouldn't be where I am today: Christy Merecka, Carlos Rodriguez, Ericka Pullin, Trey Howell, Nick Sager, Randall Forsha, Paul Northup, Ed Pearson, Glenn Acree.

—Cameron Cyr

I have a lot to be thankful for in my life, and the privilege to be able to write this book is one of those. I want to first thank my beautiful wife, Sarah, without whom none of this would be possible. While I spent evenings after work and long weekends writing, she took care of our newborn and other two kids and kept things running in the house. She has been there every step of the way and has had to tirelessly listen to all my complaining. I also want to thank my kids Zoey, Bennett, and Ellis for giving up some of their time with Dad to allow me the time and space to do this. I love you all more than anything and hope this book inspires you all to dream big.

With regard to the book itself, I cannot thank my coauthor Cameron Cyr enough. Cameron is one of the smartest people I have ever met, and his adaptability to new things, eagerness to learn, and drive to grow is awe-inspiring. He is the most knowledgeable person I have ever met in regard to dbt, and I am so thankful that he was part of this book. I have learned so much from him over the past couple years, and he has helped make me a better engineer and a better person. I hope this book is a platform that helps show the world the incredible talent and person he is.

Throughout my career, there have been a lot of people who have pushed, inspired, or provided me with the avenues to be able to write a book. Jim St. Clair, Justin Steidinger, Jon Buford, Waylon Hatch, Adam Murphy, and Randall Forsha are all former managers who gave me opportunities, challenged, and supported me in growth and learning. Also, numerous members of the community including Mark Josephson, Matt Gordon, Denis McDowell, Tim Cronin, and Ed Pearson have inspired and taught me so much over the years. And a big, special thanks to Kevin Kline, who is a great friend and mentor to me. I would not be an author without him. And lastly, thank you to everyone in the data, dbt, and SQL communities for all that you do and for letting me be a part of you.

—Dustin Dorsey

Together we want to acknowledge some people who helped make this book a reality, starting with our amazing technical reviewer, Alice Leach. When we thought about whom to reach out to about this project, Alice was one of the first people who came to mind. We met her at the Nashville Data Engineering meetups and quickly learned that she was as passionate about dbt as we are and very knowledgeable about the subject. Her comments and feedback really helped shape a lot of the content you are reading today, and we are very thankful for her involvement. We also want to thank Apress for giving us the opportunity to write for them. Thank you Jonathan Gennick and Shaul Elson for all of your involvement throughout the process and your help in making this book a reality.

Next, we want to thank dbt Labs and their team for the incredible product that they produced. Without them, there is no content or book to write. We truly believe dbt is one of the greatest data tools created in recent memory and is one that will be used for years and years to come. Also, we want to thank the dbt community, and particularly those in the dbt Slack channels, who are always willing to share their experiences and knowledge and answer questions. You all are amazing and continue to be our go-to source for anything dbt related.

—Cameron and Dustin

Preface

In this book, we embark on a journey to explore the powerful world of dbt and its transformative potential in the realm of data and analytics. Whether you are a seasoned data professional or someone just starting to dip their toes into the vast ocean of data, this book aims to equip you with the knowledge and skills necessary to leverage dbt effectively.

Data is the lifeblood of modern organizations, driving informed decision-making and enabling businesses to stay competitive in a rapidly evolving landscape. However, the process of transforming raw data into actionable insights is often complex and time-consuming. This is where dbt comes into play: it provides a robust framework for managing and executing data transformations in a more efficient, scalable, and collaborative manner. While there are many tools and services available that can do this, none utilize the foundational skill of writing SQL as effectively as dbt does.

Our primary objective with this book is to demystify dbt and empower you to unlock its full potential. While growing astronomically, the product is still relatively new as far as technology goes, so finding information can sometimes be difficult. We feel like some areas of content creation are still growing and working to catch up, especially as it relates to getting started or building understanding around everything the product is capable of. When we started writing this book, there were no books on dbt, sparse blogs and video creators, and limited information outside of the content dbt Labs produces. This book was our attempt to change this and provide users of dbt with a new resource to help them on their journey.

Whether you are looking to design data models, orchestrate complex workflows, or automate data pipelines, this book serves as your comprehensive guide. Through a combination of practical examples, best practices, and real-world use cases, we aim to make the learning process engaging, interactive, and applicable to your own data projects. This book is not a rehash of vendor documentation and training videos, but it is based on our real-life experiences of using dbt to build production enterprise data warehouses that are powering real businesses today.

Chapter by chapter, we delve into the key concepts and techniques that underpin dbt and use them to continually build on each other. We start by laying a strong foundation, covering the core principles and architecture of dbt. We then move on to exploring the different components and features of dbt, including setting up your project, building models, testing, documentation, and deployments. Along the way, we address common challenges and provide insights into optimizing performance, ensuring data quality, and promoting collaboration within your teams.

Moreover, this book acknowledges the growing trend of cloud data warehouses and their integration with dbt. We delve into the nuances of deploying dbt on popular cloud platforms such as Databricks, Amazon Redshift, Google BigQuery, and Snowflake. With step-by-step instructions, we guide you through the setup and configuration process, enabling you to harness the full potential of dbt within your cloud environment. All examples throughout the book were created using a Snowflake instance but are simple enough that most will work with any cloud data warehouse with minimal or no changes. We do not seek to push you in a certain direction, but instead provide you with the know-how to enable you on whatever direction you choose.

Throughout this book, we aim to foster a sense of curiosity, experimentation, and continuous learning. The field of Data Engineering and Analytics is constantly evolving, and dbt is at the forefront of this revolution. By the time you reach the final page, we hope that you will have acquired the necessary knowledge and confidence to embark on your own dbt journey, unlocking the true potential of your data and transforming the way your organization operates.

So, let's embark on this exciting adventure together and dive into the world of dbt. Let's unlock the transformative power of data and revolutionize the way we build, analyze, and leverage it. The possibilities are endless, and the time to start is now!

CHAPTER 1

Introduction to dbt

In 2006, Clive Humby, a British mathematician and entrepreneur in the field of data science, coined the phrase "Data is the new oil" because of its incredible high value. While both are valuable in their raw state, they must first be processed in order to create something of value. Crude oil is sent to refineries where it is separated into petroleum products like fuel, oils, asphalts, etc. which turn it into staples of everyday life. Data is refined by people, tools, and services into reports, data warehouses, machine learning models, etc. to extract its value. Take both data and oil and leave it in its raw form and what you have is a wasted opportunity.

For as long as we can remember and as far as we can see into the future, data is always going to be important. For many companies, their data is estimated to be worth more than the company itself. It is also why it is not only one of the hottest fields in tech but one of the hottest fields in any industry. Demand for skilled workers has exceeded the available talent pool which has left some companies trying to build their own talent or just scaling back important initiatives. The data tools and services that are thriving in today's market are ones that play to the strengths of most data professionals and not force learning an entire new programming language, especially when learning a new programming language does not offer substantial advantages.

This is what we as the authors of this book love about dbt. One of the core skills any data professional needs to have is writing SQL, and dbt makes that the core of what it does. So while it is a relatively new tool, the learning curve for any data professional comfortable writing SQL is small. This chapter will serve as an introduction to the rest of the book. This will cover what dbt is, who it is for, how it fits into your data stack, and more. We will also do a high-level overview of the product to introduce fundamental terminology that will be used throughout the book. There will be a lot covered in this chapter, so don't worry if you feel like you are lost at the end because the rest of the book will dive into specific topics related to dbt so that you will learn everything you need to know to get started using it. We want to provide you with a glimpse of the big picture and then take it step by step the rest of the way to understand each component.

C. Cyr and D. Dorsey, *Unlocking dbt*, https://doi.org/10.1007/978-1-4842-9703-2_1

What Is dbt?

At its simplest definition, dbt (or data build tool) is an open source data transformation tool commonly used for modeling data in data warehouses. It enables data professionals comfortable writing SQL to transform data in their warehouses more effectively. At a more granular level, it is a tool that has a development framework that combines modular SQL with software engineering best practices to create a better, faster, and more reliable data transformation experience. Modular SQL refers to the practice of breaking down SQL code into smaller, reusable chunks so that it can be more easily maintained, tested, and reused. Other best practices that dbt lends itself well to include version control, documentation, DevOps, and so on.

It is very important to understand though that dbt is just a transformation tool and does not handle data movement outside the data store it connects to. So when you consider extract-transform-load (ETL) and extract-load-transform (ELT) designs, this tool is only the transform component of these patterns. You will need to combine dbt with other tools to create your full ETL/ELT pipelines. Several market tools such as Fivetran, Airflow, and Meltano allow you to run dbt as part of their framework, or you could utilize completely separate tools entirely. How you utilize it is up to you and what works best for your use case.

Note dbt is just a data transformation tool and will need to be combined with other tools/services to create complete ETL/ELT pipelines.

Most Data Engineering tools/services are built like software engineering tools, so many of these best practices are native to them. But for Data Analysts, and their common workflows, this has not always been the case. We know job titles can have different meanings for different people, so to clarify what we mean when we say a Data Analyst, it is someone whose job is to gather and interpret data in order to solve a specific problem for the business. As the title suggests, they are responsible for analyzing data. Often, they may be responsible for also creating reports, but that is just part of the role. A lot of their work is spent writing SQL queries to understand and model the data to answer specific questions. The Data Analyst could use views and/or stored procedures to contain their code, but even still you need other tools combined with it that are often more geared toward SQL development. Additionally, creating objects on a database can be closely guarded by administrators and often require proper controls to be in place. This area of working with data has been neglected for years, and this was the primary reason dbt was originally created.

dbt was created in 2016 by a company called Fishtown Analytics that would later be rebranded to dbt Labs in 2020. Fishtown Analytics, an analytics consulting company based in the Fishtown neighborhood of Philadelphia, was founded to help series A- and B-round venture-funded companies implement advanced analytics. Specifically, they wanted to improve the way that Data Analysts do their jobs. They believed that data analysis should be practiced in a way that closely resembles software engineering and sought to find a solution for this. Since there was a gap in this market, they created dbt to help fill the void. Interestingly enough, dbt was originally an open source solution to add basic transformation capabilities to Stitch, but the vision for it was much bigger than that. Fishtown Analytics (prior to the rebranding) built a paid version of the tool called Sinter that would provide users with additional functionality, including scheduling, logging, alerting, and more. In 2019, this tool was rebranded to dbt Cloud, and the functionality has skyrocketed into what it is today. Both dbt Core and dbt Cloud are still available and are subjects that will be covered in depth throughout this book.

The open source element, dbt Core, is the codebase needed to be able to operate dbt. You will need your own tools and infrastructure to be able to run this, but it is free to utilize the codebase even if other elements do generate costs. Even if you aren't using dbt Core and elect to use the paid version, dbt Cloud, then you do have some control over the codebase. There are macros and other features that you can alter that impact the default behavior, but you can't change everything like you can with core. There are other components related to dbt Cloud that are not open source though specifically related to the integrated development environment (IDE), scheduling, jobs, etc. We will cover this at a high level later in this chapter and spend a lot more detail with this in Chapter 2.

As stated previously, dbt is a data transformation tool, but it is unique to other tools that exist on the market. A couple of quick callouts about dbt that we think are important aspects you need to understand early on:

1. dbt does not contain the computation needed to execute your models and requires the compute of your warehouse/database.

2. Every model, except for Python models, that you build in dbt is written as a SQL select statement and compiled as a SQL statement regardless if you use Jinja or not.

3. dbt handles database object creation (tables, views, etc.) for you, so no need to create these ahead of time.

First, while dbt is just a transformation tool, it does not contain the compute power needed to execute statements on any of its offerings. Everything that dbt queries and builds against your warehouse utilizes the compute power from your warehouse. For example, if you are using AWS Redshift as your data warehouse or data store, then dbt will use AWS Redshift's compute cluster to build your models. dbt does not have a computing mechanism to handle data processing, and we don't foresee that changing. There is a small amount of additional compute needed to develop with and schedule dbt jobs with dbt Cloud that is included, but does not utilize your warehouse compute. However, if you are developing with dbt Core, this is something you will need to consider when designing your infrastructure. Also, as mentioned earlier, dbt does not have the ability to extract and load data, so the data must be present on the same system you are transforming it on. So if you are utilizing the compute associated with AWS Redshift, then you need to have the data available in AWS Redshift before running dbt models against it.

Second, it is important to understand that every model you create in dbt compiles as a SQL statement that gets run against your warehouse. The one exception to this is Python models, but we will cover those in detail in Chapter 4. dbt adds in Jinja to be used with your SQL that can be incredibly powerful and efficient, but it still all just translates to SQL code. We will discuss Jinja more later in the chapter, and then in depth in Chapter 6, but for now you should just be aware that Jinja is a templating language. Interestingly, every model you build must also be written as a SELECT statement, which takes some getting used to if you are used to writing your own data manipulation language (DML) statements (UPDATE, INSERT, MERGE, and DELETE) to modify existing database objects. Depending on your model type though, the dbt compiler will add any required DML or DDL (data definition language) boilerplate code needed to manage your database objects. We will cover this in depth in Chapter 4.

The last big component we want to call out here is that dbt also handles your object creation against the warehouse for you. If you have worked with data warehouses in the past, you may be used to creating your schema objects before transformations, but this is no longer necessary. While you could, we always advise against deploying this antipattern and recommend utilizing dbt to create your objects. You still have control over setting data types if you don't want to infer them, but you need to do it through your code using things like CAST commands.

The Analytics Engineer

If you spend much time around dbt, it is inevitable that you will hear the term Analytics Engineering. This is a term adopted by the dbt team to label what users of dbt do. Analytics Engineers are the combination of what you think of when you combine a traditional Data Engineer with a Data Analyst. They serve up modeled and clean datasets to business users to consume the data. Data Analysts typically spend their time analyzing data, whereas Data Engineers spend their time building infrastructure, extracting, loading, and transforming data. Using the power of dbt, an Analytics Engineer combines elements of these and creates a bridge from Data Analytics into Data Engineering using skills they are comfortable with. Claire Carroll wrote a blog on the dbt site that gives a great overview of what Analytics Engineering is, and in it she lists her view on how Data Engineers, Data Analysts, and Analytics Engineers are different.[1] Figure 1-1 highlights some of the things she mentions in that article.

Data Engineer	Analytics Engineer	Data Analyst
• Build custom data integrations • Manage overall pipeline orchestration • Develop and deploy machine learning endpoints • Build and maintain the data platform • Data warehouse performance optimizations	• Provide clean, transformed data ready for analysis • Apply software engineering best practices to analytics code (ie. version control, testing, CICD) • Maintain data documentation and definitions • Train users on how to use data visulaization tools	• Deep insights work to answer business specific questions • Work with business users to understand data requirements • Develop and deploy machine learning endpoints • Building reports and dashboards • Adhoc data analysis

Figure 1-1. *View of an Analytics Engineer*

As authors of this book, we love the role of Analytics Engineering, but we don't feel there is anything revolutionary about it and think of it more as a great marketing strategy to provide a unique name to users of dbt. However, dbt has provided those who sit in the role of Analytics Engineer the toolset to deliver value faster. Gone are the days of having to wait for the Data Engineering team to implement a simple transformation. Job titles alongside roles and responsibilities can vary greatly between organizations, and the role of an Analytics Engineer could very easily be connected to a Data Engineer or Data Analyst, and we feel like it fits. Analytics Engineering takes aspects traditionally associated with both of these roles and combines them together. Companies (especially

[1] See `www.getdbt.com/what-is-analytics-engineering/`

ones with smaller teams) have been doing this for decades, but dbt provides a platform to make this more efficient. We are not saying this as a negative though. The new title has gained a ton of popularity, promoted brand awareness, and provided a label for dbt users to connect over which is great.

We also feel like dbt is marketed very heavily to the Data Analyst community, and it makes a lot of sense when you understand their market. dbt was originally created as a way to introduce software engineering best practices to the underrepresented Data Analyst role. While we absolutely believe that is the case, we want to state that dbt is also a Data Engineering tool and a really great one to that point. We don't feel like this gets the marketability it deserves, because it is a fantastic Data Engineering tool for anyone responsible for modeling data using SQL or Python. When considering the *why*, we believe it boils down to money and not product usability. dbt Labs earns money from subscribers to dbt Cloud, and most Data Engineers are going to be comfortable using dbt Core which is free and open source (and many do), whereas Data Analysts are less likely to be interested in managing a dbt Core deployment. While this is true, we still feel like there is some missed opportunity here. Both authors of this book fall more on the Data Engineering side of the spectrum, and dbt is one of our favorite tools to use for building data products. Even though we, and many other Data Engineers, are comfortable running dbt Core, we have spoken with many Data Engineers who also like dbt Cloud because of the ease of deployment.

Note dbt is not just a tool for Data Analysts, it is also a fantastic tool for Data Engineers.

The Role of dbt Within the Modern Data Stack

Before we understand the role of dbt, we should first understand the importance of ELT (extract-load-transform) builds. We are excluding ETL (extract-transform-load) here because most cloud architectures do not take this approach even though the following statement stands true for it as well. ELT has become common in many warehouse designs because of the power of modern analytic databases. Modern cloud data warehouses like Snowflake, Synapse, Databricks, and BigQuery are faster, more scalable, and more than capable of handling transformations at the database level. This is especially true when factoring in the separation of storage and compute with systems like Snowflake and Lakehouse architectures.

One of the main reasons that ELT models are popular with cloud services is because of network latency and the time it takes to transfer data. While network speeds are fast and only continue to get faster, data volumes keep growing as well. So if you are dealing with large volumes of data that need to be transferred over the Internet, it could still take a while. So you want to make sure you only have to load the data once. For simplification, let's say that I have ten tables that total 100GB that need to be transferred from an on-premise database server to an AWS S3 bucket, and it takes me five hours to do it. If I added transformations to it (ETL setup) and found out I did something wrong, I would have to redo it all over again. Or what if a month down the road, my requirements changed for how I was using that data, I would have to reload it again along with whatever new data was added to that size in that timeframe. Factor that over an expanding number of source systems you are pulling data from, and it doesn't take long to find yourself wasting a bunch of time. If we had just loaded the raw data (ELT), we would have only needed to load the data once. Yes, you are effectively storing an additional copy of your source data, and transformation on ingestion could narrow that, but storage is so cheap in the cloud that for most this should never be an issue. With the raw data landed, you can now work with it much faster, and if you make a mistake, you easily adjust or even start over which is a huge time saver. Additionally, as new projects come along that also need to utilize the data, you don't have to pull it again, effectively reducing your storage footprint and time.

Note dbt is not just limited to building a data warehouse and can be used to support other reporting and analytical workloads where a data warehouse may not be necessary though the data warehouse is the emphasis for this book.

So in a data warehouse ELT architecture, let's take a look at the high-level steps associated with this and how dbt accomplishes this. Figure 1-2 shows a diagram of typical items found in a data warehouse build. Regardless if you are using dbt as your transformation tool, these are still things that you should be looking at in whatever your tool of choice is.

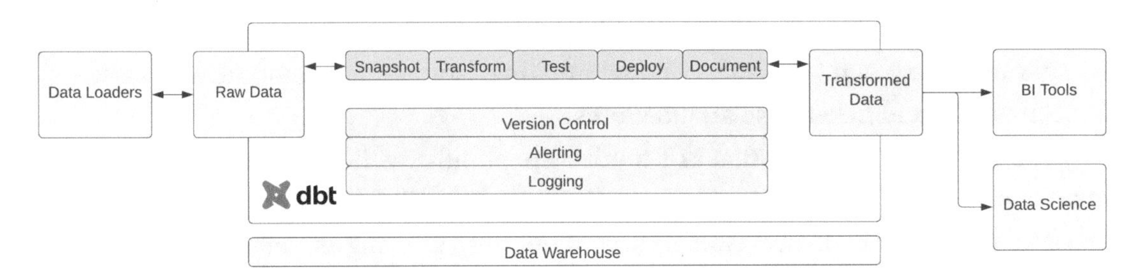

Figure 1-2. *High-level steps for an ELT data warehouse design*

The first step in building your ELT warehouse is defining and configuring your **data loader**. Your data loader is what will be used for developing your ingestion pipelines to get data from your sources into your warehouse environment and could be anything you choose. As we've mentioned before, dbt is only a data transformation tool, so this component is outside the scope of its use; however, it is still a very important component needed before you can utilize dbt. If you aren't sure where to start or what to select, we recommend connecting with the vendor you are using for your data warehouse and get their feedback. Some ingestion tools integrate really well into their system and could offer additional advantages, while others may not work very well at all. Additionally, all of the major cloud providers have their own tools you can use for this (Azure Data Factory, AWS Glue, Google Cloud Data Fusion, etc.), but don't sleep on third-party tools because there are some really great and easy-to-use ones on the market. Some tools like Fivetran have native connectors that seamlessly replicate your source data to your warehouse and take a lot of the development time and work out of this part of the process. Just be sure to do your due diligence here and really look at all options since this could make or break a project.

Once you have selected and set up your data loader, you are now ready to start **loading your raw data**. As we discussed earlier, it is important to ingest the raw data into your warehouse and never change it. This way, data can be shared for multiple uses, and any mistakes that are made working with the data can be easily resolved without needing to ingest the data again. Where you choose to load the data is another choice you will need to make and can vary largely depending on your architecture. Some designs first load the data into a storage account (often referred to as a data lake), and some designs may load the data directly into the data warehouse. Either is fine, but know that if you are using dbt as your data transformation tool, the raw data will need to live on the same system you are transforming the data on. So if you are loading data into a data lake first, but you are pointing dbt to an AWS Redshift instance, then you will also need to stage the data on Redshift first before you can use it. Of course, you could just copy the raw data directly in Redshift which would work, but there are other reasons, outside the scope of this book, why you may not want to do that. As authors of this book, we highly recommend evaluating lakehouse designs or Snowflake as potential solutions and just landing the raw data directly onto it. The separation of storage and compute is a huge advantage of using these types of platforms. Whatever you decide, just know that the raw data has to be stored in the same spot you plan to transform it to utilize dbt.

Now that you have the raw data landed in your data store, you are ready to start processing it. The first step in processing it is to **snapshot the raw data** (not to be confused with the snapshot features in dbt). We never want to change the raw because any mistake could cause us to need to ingest data again or could even cause irreparable harm if we are tracking history as part of the ingestion process. Instead, we can take snapshots of the data. This means we pull the data we need from the raw data objects/tables to start working with it. How you store it (table, view, common table expression (CTE), etc.) and utilize it is left up to how you design your model, so it could vary from organization to organization. This step is also not mutually exclusive and could be combined with the next action including transformations. As authors, we like to snapshot our raw data and combine initial transformations with it before loading into a staging table in our warehouse. We add incremental logic as well to make sure that we are only extracting and transforming new and changed data and not the full dataset each time, but again this is all based on how you design your warehouse. dbt makes it very easy to read data from your raw data, which will be a subject covered in detail throughout this book.

The next step is **transforming your data** to fit within your data model. This is reshaping your data by normalizing/denormalizing, cleaning, and modeling it for analytics, machine learning, etc. This is taking data from other disparate systems designed to handle data one way (i.e., transactional systems) and completely changing it to be optimized for reporting and analytics. This is the core of what dbt is and why you choose to use it. How you transform the data can vary greatly based on how you have constructed your data model. In the next section of this chapter, we will look at common data model types you could choose.

Once the data is transformed, you will want to **validate that the data is accurate by performing tests**. Accuracy is the most important component of any analytical or reporting role working with data, and this is especially true when building an enterprise data warehouse where that data is shared across your organization. Oftentimes, we are pulling data from systems, such as transactional systems, that are built for entirely different purposes, so we have to perform a lot of transformations to model and clean the data which creates opportunities for mistakes. Testing the data makes sure that you haven't inadvertently changed the meaning of it and that the data is modeled in the expected way. A lot of times testing can happen outside of the transformation process, which is not ideal since the damage is done by the time you find out about it. Fortunately, dbt incorporates testing as part of your transformation pipelines. This is a huge advantage for dbt and one we cover in great detail in Chapter 8 of this book.

The next item is **deployment**. You shouldn't be developing directly against production, so you need a way to safely, efficiently, and quickly move tested code between environments. While you could do it manually, most engineers want a way to automate it, so they often look at DevOps processes such as CI/CD (Continuous Integration/Continuous Deployment). There are several cloud-based and third-party tools that you can use for this, such as Azure DevOps in Azure, but dbt Cloud makes it easy to handle from within a single tool. In Chapter 10, we will cover how to run CI jobs in dbt Cloud and within GitHub Actions for dbt Core users.

Once your code is developed, tested, and deployed, the next stage is **documentation**. Documentation is an often overlooked component that only gets handled if it is made a priority, and frequently it isn't. Nevertheless, it's an important aspect of any successful data warehouse project. Many technical professionals complain about this part because technology changes so much that the information becomes out of date too quickly and no one uses it anyways. There are tools that help with this by automating some of the work, but often represent another service you need to pay for and manage. With dbt, you can add descriptions about your objects as you build, and dbt will combine that with metadata from your project and put it into an easy-to-understand website. The documentation gets refreshed by running a simple dbt command, so you never have to worry about manually updating documentation. As authors of this book, we love having this component built into the tool we are developing with and not having to spend so much time on upkeep of documentation. We explore documentation within dbt in detail in Chapter 9 of this book.

While we are refining our data throughout the warehouse, there are other components that we need to factor such as incorporating software engineering best practices. Within data pipelines, **we need version/source control for our code**, **alerting on our processes**, and **logging to be able to identify issues**. These are commonly found in many enterprise software engineering tools, and dbt is no exception here.

Once everything is built, you have your final **transformed** data, and you are ready to serve the data up to the business or other data teams to consume. This data could be consumed by data science teams, data analytics, reporting teams, and more.

All of these components should be factors and considerations that you are evaluating when considering what to use and how to build your data warehouse. As the old saying goes, "there are many ways to skin an eggplant" (we like cats), so there are lots of solutions you could be successful with and wouldn't dare state dbt Cloud is the only way.

Modeling Your Data

dbt is awesome and really simplifies a lot of the steps you need to do to transform your data, but just like any development tool, it can quickly become the wild, wild west if you're not strategic about how you use it. If you are using dbt to build a data warehouse or pipelines for shared purposes, you still need to give careful consideration to how you want to model your data. Not making a decision is a decision in and of itself. If you are just providing it as a tool to assist your analysts in how they store and use code, then you still need to think through things such as project structure, naming conventions, etc. But for this section, we are going to focus on the different ways you could model your data. dbt is agnostic to how you model your data, and it doesn't care how you do it. It will continue to function in the chaos even if your business can't.

Note dbt is agnostic to how you model your data, so the decision is up to you on the best way to do that and support your business.

Before starting, you should first determine if you need a data warehouse. Not every scenario will require this, and it is important to think this through before getting started if you haven't already. Here is a list of questions that may mean you would benefit from creating a data warehouse if you answer yes to any of them:

- Is it important to have a single source of truth for your data?
- Are multiple areas of the business going to be consuming this data?
- Do we need to analyze data from different sources?
- Do we need to maintain historical tracking of data and make sure we are looking at measurables from a point in time?

- Do you have a high volume of data that needs to be transformed?
- Do you find yourself running a lot of complex queries regularly?

If you answered yes to any of those questions, then you may want to consider investing time and resources into a data warehouse if you don't have one. If you make the determination of needing one, then you need to evaluate how you plan to structure your data.

One of the biggest mistakes we see made with dbt is it being handed to analysts and developers without a plan and letting them go crazy. This may be fine for ad hoc purposes where the content being produced is only relevant to the developer, but when you are building systems that are intended to be used at an enterprise level, this will create issues. Some of the most common issues are redundant work being done, multiple versions of the truth being created, inefficient queries, and difficulty of transferring knowledge to other members of the team. We believe that you still need a strategy around what you are building; dbt is great, but it isn't magic (okay it is, but not in every way). Whether it is a data warehouse or just data analytics, you need to have a plan and create structure around what you are building. If you decide that a data warehouse fits your needs, then you need to determine how you plan to approach the data model. There are a lot of approaches to data warehousing with pros/cons to each that you will need to consider when deciding which fits best based on the business need. While we will not cover how to select a data modeling approach for your data warehouse in this book, we will share a list of some of the most common model methodologies:

- **Data vault**: Utilizes hubs, links, and satellites to store data
- **Dimensional**: Utilizes facts and dimensions to store data
- **One big table**: Denormalized data where all values are stored in a single table
- **Relational**: Data stored in a normalized state
- **Common data models**: Open source and community-supported data models used within specific industries

There are others that are also out there, and several architects/data modelers/ vendors use their own variation or combination of these techniques. Ultimately, it is up to you to decide what works best for your business case. Whatever you decide is fine, just be sure you have a plan that you feel good about.

As authors, we have spent much of our careers building dimensional models, so the content and examples will steer toward that. This does not mean this is the only (or even best) approach to how you should use dbt. We are biased to it, but also recognize many teams have been very successful with other approaches. How you model your data is entirely up to you, we just recommend really taking the time to think through it in the beginning. Many companies have neglected this step and are now paying fortunes to consulting companies to come in and fix it.

The Skills Needed to Use dbt

Before starting to use dbt, it is important to understand what skills are involved with using it. This is something we look at with any new tool we are using, so we can determine how well our existing skill sets translate to it and always appreciate when it is laid out well and explained rather than leaving it to the user to try to interpret it.

dbt is just a framework to work within, and you still need other skills to be successful at using it. While dbt is a skill in and of itself, there are other aspects that really make up the landscape of what it means to be an Analytics Engineer. The good news is that if you have worked with data before, then it is not a difficult tool to learn and adapt to using. Compared to other tools on the market, it is one that in our opinion is the easiest to learn.

These are some of the core technical skills that will be helpful to have some background in prior to starting with dbt. Though not all are created equal, and many of these can be learned as you embark on your dbt journey:

- SQL (primary skill)
- Jinja (templating language for Python developers)
- YAML (configuration and property files)
- Python (depending on the use case)
- Data modeling
- Source control

In the following sections, we will review each skill and the role that it plays in dbt and provide an experience score that gives you an idea of how important having prior experience is. The scale will be a value between 1 and 5 with 1 meaning that we don't feel like you need any prior experience and 5 being that you should be very experienced.

Skill #1: SQL

The primary skill you need to be effective at dbt is **writing SQL** (structured query language). dbt only requires that you be able to write simple SELECT statements, which is one of the first things anyone learns when learning SQL. As long as you are capable of writing those, then you can get started using dbt. However, you will be a more effective dbt developer if you have advanced SQL coding skills. While simple SELECT statements can do some things, you are likely going to be able to do much more with complex queries throughout your journey. For example, we recommend you become comfortable with CTEs and window functions to take your SQL skills to the next level. SQL is the one skill out of everything on this list that we believe that you need to be comfortable with. If you aren't, then you may want to consider taking a step back to learn and familiarize yourself with it before continuing your dbt journey. As part of this book, teaching you to write SQL is not something we will be covering and recommend utilizing the multitude of other sources written about this subject.

Note Being comfortable writing SQL SELECT statements is the main technical prerequisite you need to have before starting your dbt journey. We believe that the rest are skills easily developed or not entirely necessary for most use cases.

Experience Score: 4/5, dbt does not require you to have advanced SQL knowledge to use, but it does require you to comfortably be able to write SQL. You can level up your skills here if you have foundational knowledge; however, if SQL is entirely new to you and you have other skills, we would recommend looking at other tools first that cater to those.

Skill #2: Jinja

The next important skill is Jinja. Jinja is a web templating language used in Python that opens up a lot of possibilities that are complicated to do with SQL logic or requires you to be able to write functions. Unless you have used Python before, this may be something that is new to you. Don't worry though, it is not as scary as it sounds and can be easily learned with dbt. There are a couple of core Jinja commands that you need to know, but the rest is stuff that you can level up as you get started using the product.

dbt caters really nicely to SQL developers, and we know that this skill will be new to people looking to utilize dbt, so we have dedicated Chapter 6 to exploring Jinja deeper. It is definitely something you need to learn to get the most from dbt, but you could still have a very successful project with only limited use of it.

Experience Score: 2/5, it is helpful to have had experience with Jinja (or Python), but it is not essential. The basics here are really easy, but the more advanced use cases may have a light learning curve.

Skill #3: YAML

Having a basic understanding of YAML is a good skill to get started using dbt since it is utilized to define configurations and properties about your dbt project. However, it is helpful to have some experience working with it but not something we consider a must-have as a prerequisite. One of the advantages of YAML is that it is very human readable and easy to learn, so even if you have never used it, it is not difficult to learn. We will cover YAML basics later in this chapter, and the book will provide you with a strong foundation of getting started using it. What you will find with dbt is that a lot of the YAML changes you make are repetitive, so once you find your rhythm, you should be fine.

Experience Score: 2/5, having YAML experience is definitely helpful, but we think it is simple enough that anyone can pick up on this quickly without any prior experience.

Skill #4: Python

dbt is primarily a SQL-first tool; however, it does also support running Python. We do not feel like this is a requirement when using dbt and is a nice-to-have that potentially only a select amount of users will use. As of this writing, Python support is new, so it is possible that it could become a more essential part of the dbt ecosystem in the future. However, since this feature is available and could fit some use cases, we wanted to mention it. The advantages of being able to run Python is that it can be used to write custom scripts and plug-ins that can extend beyond the functionality of SQL and can perform some operations with a lot less code. The dbt Labs team has specifically came out and said that this is not intended to replace SQL, but is there more as an option to fill the gaps the SQL can't.

This book will not teach you Python, but we will dive into how to build Python models with dbt and use cases of when you may want to consider using it with dbt in Chapter 4. The value of your experience here will largely depend on whether you intend to use Python models or not. Knowing Python is not essential to being able to use dbt, but it does help.

Experience Score: 1/5 for most, but if you know you need to utilize Python, this number would be higher.

Skill #5: Data Modeling

Knowing how to model your data and having a plan is an important skill. This is something that we referenced earlier in this chapter and something we think gets lost on many users of dbt. We don't want to repeat everything that we said earlier, but you should have a basic understanding of the different ways you can model your data before getting started using dbt and have a plan. Understanding the different pros/cons of those models will be important to helping you strategize around your model builds.

In the case of many companies, you may have architects or senior engineers that create the model and pass it off to the team, which lowers the experience barrier. We believe it helps for all team members to have a basic understanding of the model type they are working with, but it is not a barrier to entry if you have other capable members of the team.

Experience Score: 1/5, if you have other capable members of the team; 4/5, if you are the one leading the design and implementation.

Skill #6: Source Control

It is important to understand the basics of source control before using dbt because it is designed to be used in a collaborative environment and relies on source control to manage and track changes to your data transformation code. Source control is a way to track and manage changes to files, such as code or documentation. It is used in most areas of software development to allow multiple developers to work on a project at the same time while ensuring that changes are properly tracked and coordinated. Commonly, this is done by using GitHub, GitLab, Bitbucket, etc.

In dbt, source control is used to track and manage changes to your dbt project, including your models, seed data, snapshots, and test files. By using source control, you can ensure that your dbt project is versioned, which means that you can track and roll back changes as needed. This can be especially important when working on a team, as it allows you to easily coordinate changes and avoid/resolve conflicts. As authors of the book, we think it is valuable to at least have some familiarity with this process. If you don't, then that is fine since dbt makes it really easy to set up and use. If you are using dbt Cloud, then you have to set up a repo and connect it before you can even get started.

Experience Score: 2/5, you should have a basic understanding of how source control works. Any experience will translate to this.

The Benefits of dbt

The data market is flooded with tools that you can use to manage your data and build pipelines, so what makes dbt so special? There are several benefits of using dbt for data transformation tasks, as compared to other tools. Notably, we have listed some of the benefits as follows, many of which we have already touched on throughout this chapter:

1. It is designed specifically for data transformation and analytics and provides a number of features that are optimized for these tasks. For example, dbt provides dependency management, incremental model builds, and data testing, which can help you optimize the performance of your data transformation processes.

2. It is built for SQL users by SQL users, which is a widely used and well-known language for data manipulation and analysis. This makes it easy for Analysts and Data Engineers to learn and use dbt, even if they are not familiar with other programming languages.

3. dbt is open source and has a strong community of users and developers, which makes it easy to find help and resources when you are working with the tool. The dbt Slack channel is, in our opinion, one of the best in the tech community.

4. dbt is designed to be used in a collaborative environment and provides features such as version control, documentation, and testing that can help you work effectively with other analysts and data engineers on your team.

5. dbt is flexible and can be used with a wide variety of data sources and target databases. This makes it a good choice for organizations that need to integrate data from multiple sources and systems.

Overall, dbt is a powerful and effective tool for data transformation and analytics and is well suited for organizations that need to perform these tasks on a regular basis. Many of the users who decide to adopt dbt do so because of one or more of these benefits. We know we cover these throughout the text, but we wanted to highlight them in a place that is easy to read.

Connecting to Your Database

As mentioned earlier, dbt does not have a compute engine and relies on the compute from the warehouse or data store that you are connecting it to. If you do not have a target configured for dbt, then you are not going to be able to run anything. dbt can be configured to point to a variety of SQL-speaking databases, warehouses, data lakes, query engines, and analytical platforms. Depending on the connection though, they have different levels of support. Each connection falls into one of four categories: dbt Labs supported, vendor supported, community supported, and yet to be built. Figure 1-3 shows a visual of the current connectors supported. Note that this list will likely be different at the time of your reading, so please check out the official dbt docs for the latest information.

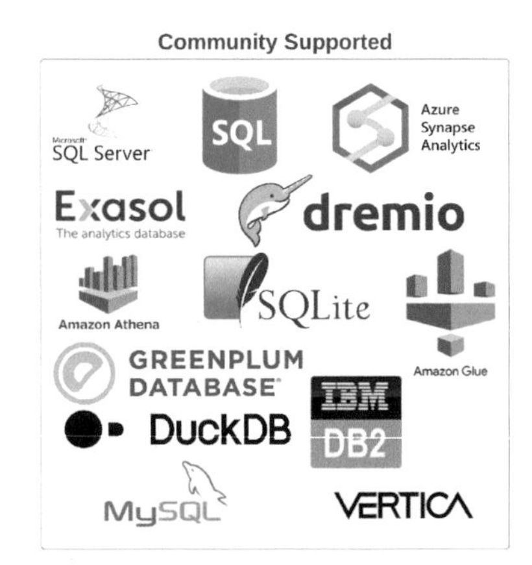

Figure 1-3. *View of supported connectors for dbt*

The first of these is **dbt Labs supported**. These are adapters maintained and supported by dbt Labs and are the only adapters currently supported with dbt Cloud. The current list of connectors includes PostgreSQL, Redshift, Snowflake, BigQuery, Apache Spark, Starburst, and Trino, though we fully anticipate that this list will continue to grow. If you intend on using dbt Cloud and don't see your platform listed as officially supported, we recommend checking the latest documentation by dbt Labs to see if it is currently out or reach out to dbt Labs support to see if it is something that is slated for release soon.

The next list of adapters is **vendor supported**. This means that these adapters are built and maintained by the product vendors, and the vendor offers support for them. These are typically stable connectors that have a sizable user base already. In order to use most of these adapters, you will need to utilize dbt Core. Today, this list includes platforms such as ClickHouse, Databricks, Cloudera Impala, Firebolt, Oracle, Trino, MindsDB, SingleStore, TiDB, Rockset, Starburst, Teradata, and others.

The third type is **community-supported adapters**. These are adapters that someone or groups of people in the community have created and shared. Some of these may work better than others, and we highly recommend testing with these prior to considering building production workloads using them. Since these are community supported, if you run into any issues you could always submit a pull request to help out with fixing issues in the adapter. There are people who have used these adapters in production environments, but we recommend using caution here as some of them may become out of date or are not always the most stable. As community-supported adapters improve, they have the ability to be verified by dbt through a rigorous validation process. Adapters that have been verified by dbt are essentially given dbt's seal of approval that they are production ready. Most of these adapters are going to be created and maintained by the vendors, but it doesn't necessarily have to be the case. If you see a missing adapter and want to build one and get it verified, then we recommend checking out the official dbt docs on this process to learn more.

The last type is **yet to be built**. dbt has the capability to be extended to any SQL-speaking database, warehouse, data lake, query engine, or analytical platform by means of an adapter plug-in. If you are interested in building your own adapter, then dbt provides instructions on how to do so. This is outside the scope of this book, and we recommend you check the official dbt documentation for more information if this is an avenue you are interested in exploring.

dbt Cloud vs. dbt Core

There are multiple ways that you can utilize dbt. One is using the free open source version called dbt Core, and the other is using the enterprise product called dbt Cloud.

dbt Core is the free open source version of dbt that can be installed and run on your own computer or server. It is designed to be used with a wide variety of data sources and target databases and provides a number of features that are optimized for data transformation and analytics. It is important to note that dbt Core does not have any compute resource associated with it directly, so you will need to utilize another service to schedule and execute your dbt jobs. We will cover the setup in dbt Core in Chapter 2; however, Figure 1-4 shows an example of what a dbt Core project looks like in the popular VS Code IDE.

Figure 1-4. *View of a project setup with dbt Core*

dbt Cloud is the paid cloud-based version of dbt that is hosted and managed by the dbt team. It provides many of the same features as dbt Core, but comes with several added components that simplify the use. These include the following:

- Browser-based IDE (hosted compute)
- Job scheduling (hosted compute)
- Logging and alerting
- GitHub and GitLab integration
- Hosted documentation
- Only dbt Labs–supported adapters

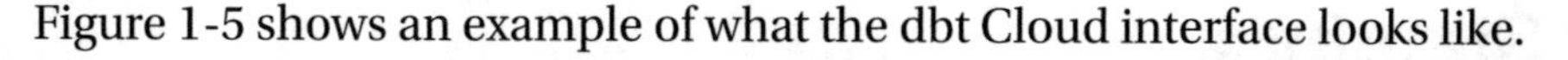

Figure 1-5 shows an example of what the dbt Cloud interface looks like.

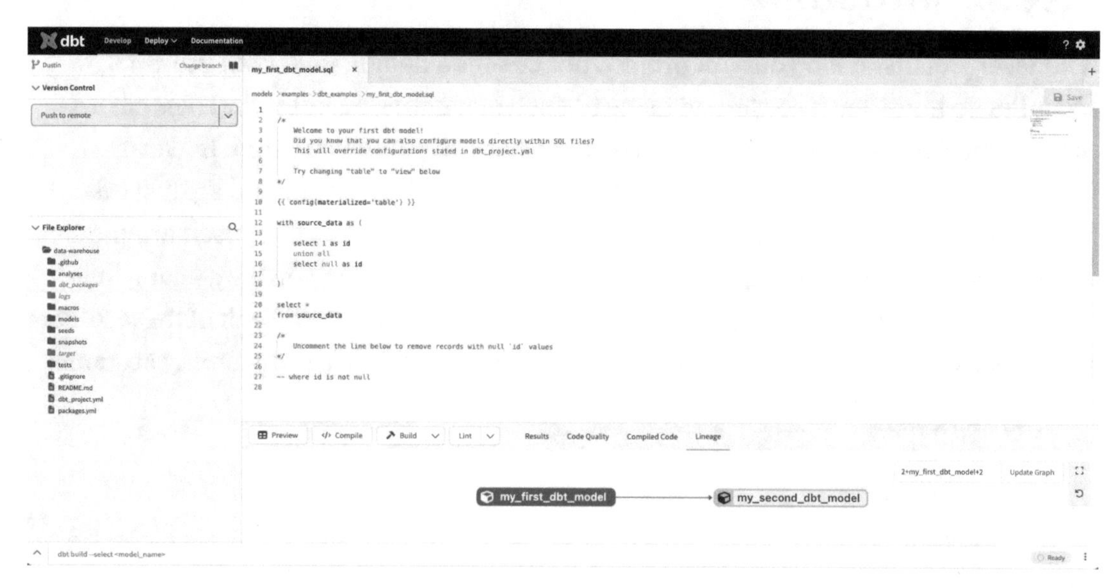

Figure 1-5. *View of a project setup with dbt Cloud*

One key difference between dbt Core and dbt Cloud is the way that they are deployed and used. dbt Core is installed and run locally, while dbt Cloud is accessed and used through a web interface. However, there are new features of dbt Cloud in development that will allow dbt Cloud users to bring their own IDE, such as VS Code. However, dbt Cloud is easier to set up and use, but may not be as flexible as dbt Core when it comes to integration with other tools or systems. With that said, dbt Labs is constantly releasing new features to further enable integrations between dbt Cloud and third-party tools. Another difference between dbt Core and dbt Cloud is the pricing model. dbt Core is open source and free to use, while dbt Cloud is a paid service that is currently priced based on the number of users who access the service. While we always anticipate a free open source version, the way dbt Cloud is priced is always subject to change. Additionally, expect the gap between features to continue to grow in the future as dbt tries to push more people toward their cloud product.

Deciding on which you prefer to use will be something you need to decide early on. In the next chapter, we will review setting up your project and will compare the two different offerings and provide more context into the decision-making process.

Project Structure

Whenever you first build your dbt project, dbt creates a default folder structure according to how they recommend using it. Each one of these folders and files created serves a specific purpose. You can choose to reshape the folder structure however you like by editing the *dbt_project.yml* file, but we generally recommend against this, especially for beginner users. All of the documentation that exists is focused around this folder structure, and changing it could result in difficulties when you are seeking help.

With the default project build, the following folders are created. Each of these folders represents a different function of dbt, and we will cover each in the upcoming sections:

- analyses
- dbt_packages
- logs
- macros
- models
- seeds
- snapshots
- target
- tests

Figure 1-6 shows an example of what the project looks like within dbt Cloud, but the view will be very similar if you are using dbt Core through your preferred IDE (i.e., VS Code).

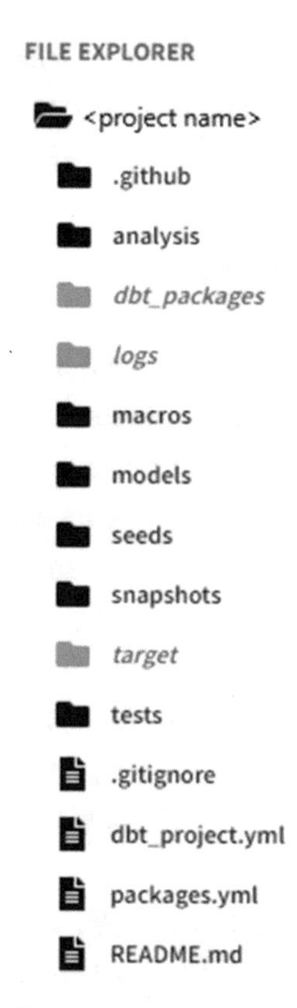

Figure 1-6. *View of the default dbt project folder structure*

The book is arranged in a way that each chapter focuses on the purposes of each of these folders, but we want to introduce it to you here so you can get an idea of the big picture. The next several sections will provide you an overview of each of the directory folders and their general purpose. We will look at them in alphabetical order, not necessarily the order in which you would build models in your project. While you can change the names of these folders and structure, we will not be doing that anywhere in the book.

Directory #1: Analyses

The Analyses folder is a place where you can store things that need to be saved, but aren't necessarily part of your data model. dbt compiles everything that is stored in this directory, but it does not execute as part of your run or build. We view this as more of a sandbox area where Analysts can archive, create, and share queries used outside your core model, but still take advantage of source control and referencing other models. Since this directory is meant to serve as a sandbox, we won't cover it any further in the book, but we did want you to be aware that it is available for you to use.

Directory #2: dbt_packages

As part of every dbt project, you have a *packages.yml* file that you will use to include other open source and community-built projects in your project. These can be hugely advantageous to you because you are able to utilize the work of other developers over having to build your own. Think of this similarly as using pip to install and use packages as part of a Python environment. There are packages that contain macros (a.k.a. functions) to simplify transformations that may require complicated SQL, and there are packages built for specific datasets that model the data for you. We highly recommend familiarizing yourself with the packages that are out there and will even recommend several throughout this book that we consider must-haves. You can check GitHub for them, the dbt Slack community, or you can check `https://hub.getdbt.com/` for the list acknowledged by dbt Labs.

The dbt_packages folder is where these packages materialize in your project. You will need to add an entry into your *packages.yml* file to include the package and then use the **dbt deps** command to materialize it. Once you do that, then models built using the package will become usable in your project. We cover setting up packages in Chapter 6, but we reference important ones you may want to consider using throughout the entire book.

Directory #3: Logs

You probably could guess this, but the logs folder is a directory within your dbt project that is used to store log files. Log files contain information about the progress and results of your dbt runs, including any errors or warning messages that are generated and can be helpful when troubleshooting failures. In most cases, if you are using dbt Cloud, you will utilize the output results generated through the integrated development environment (IDE), but they are written to this directory.

This is not something we have a dedicated chapter for in the book, but will reference this location when reviewing other subjects. We believe it is important for new users to be aware of this as it is important to troubleshoot issues.

Directory #4: Macros

Macros are pieces of code that can be reused multiple times throughout your project. They are synonymous with functions in other programming languages and are written using Jinja. In fact, the place where you will use Jinja the most is probably here. This is why we combined the topics of Jinja and macros into the same chapter which is covered in Chapter 6.

Directory #5: Models

The models folder is the main location for your data transformations in your dbt project and will be where you spend most of your time. Inside this folder, you will create single files that contain logic as part of the process to transform raw data into your final data state. You will usually have several subdirectories that exist within this folder that could include parts of the transformation process such as staging, marts, and intermediate. Having a good strategy for how you organize this folder will help you along your dbt journey. We will share our recommended design structure for this in the next chapter which discusses setting up your project. We will also dive into this topic a lot deeper in Chapter 4 of the book.

Directory #6: Seeds

The seed folder is a directory within your dbt project that contains seed data. Seed data is data that can be loaded into your target database before dbt runs any of your models. This data is typically used to populate lookup tables or to provide reference data for your analytics. It should only be used for small sets of static data that may not live in a table in a source database, but still needs to be referenced. For example, you may have a CSV file that contains a list of countries and their country code that you need to reference throughout your models. This list will be small and unlikely to change very often, so you could store it in a seed file rather than your source database. Seeds are covered in a lot more detail in Chapter 3 of this book.

Directory #7: Snapshots

The snapshot folder is a directory within your dbt project that is used to store snapshots. Because of its uniqueness, it is the only type of dbt transformation that is broken out into its own folder. Snapshot data is a copy of the data in your source database at a specific point in time. Snapshots are a feature of dbt that are used to track changes to your data over time and to ensure that your models are based on accurate and up-to-date data. Snapshots very closely mirror a Type-2 slowly changing dimension where you have valid_from and valid_to dates. This allows you to retain history of records and capture point-in-time data values. Snapshots are covered in more depth in Chapter 5 of the book.

Directory #8: Target

The target folder serves a few purposes, but the main two are to store metadata files and compiled SQL. With nearly every command execution in dbt, it will generate and save one or more artifacts that are also contained in the target directory. Several of these are JSON files (manifest.json, catalog.json, run_results.json) that are for a lot of different purposes. These include generating documentation, performing state comparisons, change tracking, project monitoring, and so much more. While we do not have a dedicated chapter to the target directory, we do reference these JSON files that are produced throughout the book and describe in more detail how they are used.

First, the target directory stores a set of .json files that contain metadata about your dbt project. The first file to be aware of is the *manifest.json* which contains a substantial amount of metadata about your dbt assets including configs, node relationships, and documentation blocks. You will never need to edit this file directly because it is generated and maintained by dbt automatically, but as you explore this file, you will quickly realize that there is an invaluable amount of metadata in this file. You can interact with this file directly to create logs, test coverage reports, and more. Behind the scenes, dbt uses this file to help generate your documentation site and for state comparison. Next is the *run_results.json* file which, as you probably guessed, contains metadata about the results of an invocation of your dbt project. It is quite common to use this file to create logs in your database so that you can monitor job results over time. Fortunately, there are packages that do the heavy lifting here for you, and we will discuss this later in the book. The last two files, *catalog.json* and *index.html*, that get generated are used solely for generating the documentation site. You won't need to interact with these directly, but they do need to be present in your project directory for your documentation site to work correctly.

In addition to these artifact files that get generated by dbt, there are two subdirectories that you will find here. The first is the compiled folder, which is used to store .sql files representing the compiled SQL of your model files. Effectively, the compiled SQL is just your model file with the Jinja stripped out and replaced with raw SQL. However, at this point, compiled SQL still is just a representation of SELECT statements, and there is now DDL or DML included yet. That brings us to the run folder, which is where the queries that are shipped to your data platform get stored. These files are where you will find SQL that includes all of the boilerplate DDL and DML that is needed to build objects in your database. Again, you won't need to interact with these directly very frequently except for when you are doing some serious debugging.

It's important to note that the target folder is not used to store seed data or snapshot data. Seed data is loaded into your target database before your models are built, and snapshot data is used as a reference by dbt to track changes to your source data. The target folder is used exclusively to store the results of your model builds and store compiled SQL.

Directory #9: Tests

The tests folder is a directory within your dbt project that is used to store test files. Test files contain SQL queries that are used to test the accuracy and integrity of your model results. dbt has several common tests that are native to the product and just require you to reference them in your YAML files; however, you can also build custom tests to support your specific business case. We cover tests more in depth in Chapter 8 of the book. There we will look at the built-in tests, custom tests you can build in this directory, and tests that are part of open source dbt packages.

Supported File Extensions

In the previous section, we covered the directories that dbt creates by default, but that is just the beginning. To utilize dbt, you will need to create files within the context of those folders that contain the programming text. dbt supports a lot of different file types, but the following are the most commonly used and ones that we show examples of throughout the book:

1. **.sql**: This is the most common file extension used by dbt, and it is used to store SQL scripts.
2. **.yml/.yaml**: This file extension is used to store configuration files in YAML format. dbt uses these files to define the structure of your data model, as well as any dependencies or transformations that need to be applied.
3. **.csv**: This file extension is used to store data in comma-separated values (CSV) format. dbt can use CSV files as seed data for your data models.
4. **.py**: This file extension is used to store Python scripts. dbt allows you to write custom Python code to perform advanced data transformations or to extend the functionality of dbt in other ways.
5. **.md**: This file extension is used to store Markdown files. You can include documentation within your dbt repository, and Markdown is a common format used for writing reusable documentation via doc blocks.

This represents what file types you can include in your dbt project today, but it may be capable of running other file types and extensions in the future as well. You can extend the functionality of dbt Core by writing custom code or using third-party libraries. This can allow you to use additional file types or extensions that are not supported by dbt out of the box. However, this will require significant additional setup and configuration, and you will need to write custom code to integrate the new file type or extension with dbt. If there is a file type extension not listed earlier that you are interested in, we recommend checking the official dbt documentation to see if it is supported. If not, then there may be solutions created by the community that you can use. For the average user, just creating .sql and .yml files are going to be sufficient.

Types of Models

Building models in dbt is where you will spend most of your time. Models are SELECT statements defined as .sql files, or defined as Python models, that are used to perform your data transformations. These are the items that make your project come to life. Within dbt, you have four different ways you can materialize a model. We will cover this in more detail in Chapter 4, but we wanted to go ahead and introduce these at a high level since it is so important in foundationally understanding dbt. The four different types are as follows:

- Table
- View
- Ephemeral
- Incremental

If you have worked with databases before, then the first two items should be very familiar to you. **Tables** are database objects where the data is written to disk and is a new self-contained object. A **view** is a database object that stores a query that is then run each time you execute the view. Data is not moved or written with a view, so the code you wrote has to be executed every time the view is. Views are also the default materialization within dbt. If you utilize either of these materializations, then the object is recreated from scratch each time you invoke dbt. For views, this is usually fine, but for tables this means all of your data has to be reloaded as well. In Chapter 4, we provide a lot of guidance on when to use one over the other.

The next type is an **ephemeral** model. Ephemeral models are essentially just CTEs (common table expressions) in SQL. They do not exist anywhere on the database and can only be referenced within dbt. We recommend using these very sparingly within your project.

Pro Tip You will want to get very comfortable writing CTEs because you will use them a lot with dbt. Because you are only allowed to write SELECT statements, you cannot use temp tables or table variables.

The next type is an **incremental** model. Incremental models materialize as a table; however, they are configured to allow dbt to insert or update records since the last time dbt ran. This is useful when you are working with large volumes of data and you don't want to have to reload it every time. You are still limited to only writing SELECT statements; however, dbt compiles that into the commands that need to be run. In other systems, you would need to write the UPSERT or MERGE logic, but dbt makes it simple for you. Working with incremental models is slightly more complicated than tables, but one you will surely use working with bigger data. dbt officially recommends starting with table materializations first and moving to incremental models when performance dictates it and we agree. Keep it as simple as possible and evolve as needed.

Snapshots

Snapshots are sort of their own thing in dbt, but similar in ways to the materializations we just reviewed. Snapshots are materialized as a table and used when you need to track history over time. Essentially, these transformations implement a Type-2 slowly changing dimension (SCD). That means whenever a change is detected between the source and the target, a new row is created in the target rather than just overriding the existing value. As seen in the project structure overview earlier in this chapter, snapshots live separately from everything else since they have some unique qualities about them. Like incremental models, we don't recommend starting with these out of the gate if you are new to dbt and only recommend starting to use these as needs dictate it. Chapter 5 is an entire chapter dedicated to walking through this concept in great detail.

Executing dbt Commands

Building models in dbt is the first step in utilizing dbt, but you still need a way to execute those models. dbt has its own set of commands that you can run to execute different tasks via the command line, including building models, running tests, and generating documentation. In the following sections, we will review all of the dbt commands that are currently supported. We have split them up based on whether they are supported by both dbt Cloud and dbt Core and ones that are only supported by dbt Core. There are currently no commands that are only supported by dbt Cloud.

Supported by Both

The following are a list of commands that are supported by both dbt Cloud and dbt Core:

> **dbt run**: This command is used to run a specific model or set of models. It is useful when you want to build and deploy only a specific set of models, rather than all of the models in your dbt project.

dbt build: This command is used to build and deploy your data transformation models. When you run the build command, dbt will execute the SQL code in your models and load the results into your target database. This command runs models, tests, snapshots, and seeds with a single command.

dbt deps: This command is used to pull the specified version of the packages stored in your *packages.yml* file.

dbt test: This command is used to run tests on your models and data to ensure that they are accurate and reliable. Tests are defined in SQL files that are placed in the test folder within your dbt project. When you run the test command, dbt will execute the tests and report the results.

dbt run-operation: This command is used to run a macro.

dbt snapshot: This command is used to take a snapshot of your source data. Snapshot data is used by dbt to track changes to your source data over time and ensure that your models are based on accurate and up-to-date data.

dbt seed: This command is used to load seed data into your target database. Seed data is a set of reference data that is loaded into your database before your models are built. It is typically used to populate lookup tables or provide reference data for analytics.

dbt show: This command allows you to preview results of a model, test, analysis, or SQL query directly from the terminal window.

dbt source freshness: This command is used to evaluate whether your source tables are stale or not based on configurations in your sources.yml file.

dbt Core Commands

The following are the list of commands that are only supported by dbt Core:

dbt clean: This command is used to clean up the target folder and dbt_project directory. It is useful when you want to start fresh with a new build.

dbt debug: This is used to test the database connection and return information that can be used for debugging.

dbt init: This command initializes a new dbt project when working in dbt Core.

dbt list: This command is used to list resources defined in your project.

dbt parse: This command is used to parse and validate the contents of your project. If you have syntax errors, the command will fail.

dbt docs generate: This command is used to generate documentation for your dbt project. It creates a set of HTML files that contain information about your models, including their structure, dependencies, and source data.

dbt docs serve: This command is used to serve your documentation locally and open the documentation site using your default browser.

dbt Command Flags

The previous section provided you with the base commands to execute aspects of your dbt project, and each of these has several additional flags associated with them. In dbt, flags are command-line arguments that you can use to modify the behavior of dbt commands. While we won't explore every possible flag for every command here, we do want to give you an idea of what is possible. One of the most widely used commands you will use is the dbt build command. You can just run dbt build, and it will run all of your transformations, tests, and seeds, but what if you want to limit the scope so it only runs a portion of the transformations? In that case, you will need to include flags to let dbt know this. Here are some of the flags associated with a dbt build command:

--**select**: This is used to only run specific transformation models and requires the name of the model following this option. This is a very powerful flag that allows you to filter based on a package name, model names, directory, path, tags, configs, test types, and test names.

--**exclude**: This is used to specify a set of models that should be excluded from the build. It can be used in conjunction with the --select flag to build all models except the ones specified. It allows the same filters as the --select flag.

--**resource-type**: This flag allows you to select the resource type you want to run.

--**defer**: This flag allows you to run models or tests in your project and skip upstream dependencies. This is useful when you want to test a small number of models in a large project and don't want to spend time and compute everything to run.

--**selector**: You can write resource selectors in YAML, save them to your project, and reference them with this flag in your command.

In addition to those flags, there are also additional flags that can be used that are not specific to resource builds. These are the following:

--**help**: This shows all of the available commands and arguments. This is helpful when you may not remember what all the options are.

--**vars**: This flag is used to supply variables to your project.

--**full-refresh**: This is used to force a full refresh of your incremental models. This will drop existing tables and recreate them.

--**fail-fast**: This flag is used to make dbt exit immediately if a single resource fails to build.

--**threads**: This is used to specify the number of threads that dbt should use to build your models. It can be used to optimize the performance of the build process by parallelizing the execution of your models.

--**target**: This is used to specify a target connection to your database. This flag is only used in dbt Core deployments.

This is just to give you an idea of what is possible with the dbt-specific commands. Each of the commands may have their own unique set of options as well, so you will want to check out the official documentation to see what is possible. Throughout this book, we will reference these commands often and may use them with some of the most common options; however, we cannot cover every possible scenario.

Case Sensitivity

When working with dbt, there is some trickiness when it comes to case sensitivity where some elements are case sensitive, some are not, and some may be dependent on the case sensitivity of your target data stores. As authors of this book, we have worked with a wide range of database systems and know this can get very frustrating especially when you have been working with case-insensitive systems, so we wanted to make note of it.

In general, dbt itself is case sensitive when it comes to compiling SQL scripts, configuration files, or other types of files. For example, the following SQL statements are **not** equivalent and will be treated uniquely by dbt:

```
SELECT * FROM {{ ref('customers') }};
SELECT * FROM {{ ref('Customers') }};
SELECT * FROM {{ ref('cUsToMeRs') }};
```

However, a problem can develop when the underlying data store is case sensitive. Some databases, such as PostgreSQL, are case sensitive by default, which means that they will treat uppercase and lowercase characters differently. In these cases, dbt will respect the case sensitivity of the underlying DBMS, and you will need to be mindful of the case of your SQL statements and other files. Using the preceding example may produce different results with these types of systems. So you will want to make sure you understand the case sensitivity of your data store prior to getting started with dbt to save yourself some headaches.

It's also worth noting that most of what you do in dbt is case sensitive. For example, the .yml file extension is typically used for configuration files written in YAML format, and YAML is case sensitive. This means that you should be careful to use the correct case when writing YAML configuration files, as dbt will not automatically correct any case errors.

When running dbt commands from the command line, dbt is also sensitive, just like it is when processing SQL scripts and other types of files. This means that you cannot use uppercase, lowercase, or a combination of both when typing dbt commands. As a rule, all dbt execution commands should be lowercase. Let's take a look at the dbt run command and see the correct casing vs. incorrect ones. The incorrect ones will produce an error in dbt rather than running:

```
dbt run - Correct
dbt Run - Incorrect
dbt rUn - Incorrect
```

dbt flags also accept arguments that are case sensitive. For example, the --select flag, which can be used to run select models, is case sensitive. This means that you will need to use the correct case when specifying the model you want to run, as dbt will not automatically correct any case errors. Let's assume we have a model called Customers in our dbt project. If I run the following command, dbt completes the transformation for this model:

```
dbt run --select Customers
```

However, if the casing does not match the name of the model exactly, then dbt will fail stating that it cannot find it and will only run once you have entered it with the correct casing.

Pro Tip Dealing with case sensitivity in dbt can be a pain, so we recommend assuming everything you do needs to be case sensitive rather than trying to keep track.

What Is YAML?

YAML plays an important part in your dbt journey, and it is crucial that you understand it. As mentioned earlier, YAML is a human-readable data serialization language most commonly used for configuration files. The meaning of the acronym can vary based on who you ask and can mean YAML Ain't Markup Language or Yet Another Markup Language. Its object serialization makes it a viable replacement over some of the complexities of working with JSON or XML.

Numerous modern tools today rely on YAML as part of its framework. Let's look at some of the benefits of using YAML:

- Human readable
- Unambiguous
- Portable across most programming languages
- Friendly to version control
- Strict syntax requirements, so everything is written similarly
- Simple and clean syntax
- Quick to implement
- Secure implementation that closes exploits found in other languages
- Easy to learn

Within dbt, YAML files are used for both **properties** and **configurations**. Properties generally declare things about your project resources and can be things like descriptions, tests, and sources. Configuration files tell dbt how to build resources in your warehouse and contain things like materializations, object locations, and tags. These will all be covered throughout the book, so if you aren't sure what they mean right now, that is okay. The key point here is understanding YAML files are used to define properties and configuration.

If you are new to YAML files, then it will help to learn more about some of the fundamentals. This is not a how-to YAML book, but an important concept to get started with YAML files in the context of dbt is to understand what arrays and dictionaries are. A YAML array is a list of similar values nested under a key that's represented by a name. Each value starts with a hyphen symbol followed by a space. Let's take a look at an example YAML file in Figure 1-7 to see how this works.

```
key1:
  - value 1
  - value 2
  - value 3
  - value 4
```

Figure 1-7. *Example array in a YAML file*

The next component is a **dictionary**. Dictionaries are mappings that are used to map key-value pairs. In YAML, dictionaries can be represented using **flow mapping** and **block mapping** syntaxes. In a block mapping dictionary, the key and value pairs are represented using a colon syntax. Let's take a look at an example in Figure 1-8 to see how a connection string could be represented using a block mapping dictionary.

```
MyDatabase:
  hostname: localhost
  port: 1433
  username: admin
  password: Password1
```

Figure 1-8. *Example block mapping dictionary in a YAML file*

In a flow mapping dictionary, the key and value pairs are separated by a comma, and entire pairs are enclosed in {} characters. So taking the same values, you can see in Figure 1-9 how they are displayed differently. In most YAML files conforming to dbt standards, you will need to use block mapping for the files to be parsed correctly.

```
MyDatabase: {hostname: localhost ,port: 1433, username:admin, password: Password1}
```

Figure 1-9. *Example flow mapping dictionary in a YAML file*

The last thing that we want to point out about YAML files is the syntax is very specific, but this serves as both a blessing and curse at times. The blessing is that since it is very strict, you don't end up with folks adding their own flavor. Anyone modifying a YAML file is going to have to use the same syntax. The curse to this is that it is so specific that you will find simple things like incorrect indentation will cause it to fail. This is a common thing and a mistake we will make when first starting to use YAML. We feel like learning YAML as you work with dbt is easy enough; however, you may want to check out some online tutorials and resources if you find yourself struggling with this throughout your dbt journey.

The Role of YAML with dbt

In the previous section, we introduced you to some of the basics of YAML and how it is generally used, but it plays a very important role in how you use dbt. Notably, it is used to define the configuration for your dbt project, including details about your data sources, target databases, models, and other project-specific settings.

In your dbt project, you will have two types of YAML files that are created. The first of these is a **configuration** type YAML file. This type of YAML file tells dbt how to operate. There is only one configuration type file in your project, and it is the **dbt_project.yml** file. This file is created automatically at the root of your dbt project anytime you initialize a dbt project and must always exist for dbt to run. This is also a one to one with your dbt project, and you cannot create multiple copies. This file tells dbt how to build resources in your warehouse and includes things such as default folder structures, materializations, object locations, and tagging. If you want to change the behavior of how dbt operates, you will need to modify this file. Configurations you may consider for this file will be covered in the next chapter.

The next type of YAML file that gets created is **property** type files. These YAML files generally declare things about your project resources such as descriptions, tests, sources, and exposures. With these, you can have multiples and it is generally recommended that you do. They will also usually live in your model directories. These will be covered throughout the book in several chapters since they are used for different aspects of the project. Notably, they will be reviewed in Chapters 3, 4, 8, and 9 since these YAML files are used in those.

If you have never worked with YAML before, it is okay and you will become an expert before you know it. A lot of the way you use them is repetitive, so once you have your rhythm down, you are in a great shape.

The Semantic Layer

As you read through the chapter list and description of the book, you may notice that there is glaring omission, the semantic layer. For those unfamiliar, the dbt Semantic Layer went into public preview in October 2022 and generated a ton of immediate buzz and excitement. The function was to provide a way for data teams to centrally define business metrics upstream in the modeling layer and flow downstream to business intelligence, reporting, and data management tools - something that is a growing trend in the industry. However, upon release there were some limitations that prevented data teams from being able to use it - most notably, join navigation and limited data platform support.

In early 2023, dbt announced the acquisition of a company called Transform. Transform was a company founded by former Airbnb Engineers and largely identified as the innovators behind the semantic layer. This was a huge announcement because several of the complicated issues that dbt needed to resolve had already been done. So dbt immediately went to work on trying to integrate their semantic layer functionality, and this meant a rewrite of the existing semantic layer.

While we could write about what exists today, we don't think it is valuable. dbt has been very vocal that this will change and be re-released in late 2023, which is after this book is slated to release. By the time you were reading this, our information would be out of date, so we made the decision to just exclude it. If we have the opportunity to write a version 2 of this book in the future, this will definitely be an addition that we make.

Setting the Stage for the Book

Throughout the rest of the book, we will be diving into the specific components that have been described in this chapter. To help you follow along, a lot of the sample code contained in this book is part of an open Apress GitHub repo that you can follow along with or use for yourself. That GitHub repo can be found at `http://github.com/apress/unlocking-dbt`.

The project that we built was connected to a Snowflake instance; however, you can connect to any database you like. The code samples will be simple enough that they should work with any database with potentially only minor changes. There are a few exceptions to this, but we will mention them as the book progresses. If you want to test things out using a Snowflake instance, then you can set up a trial account at Snowflake.com, or you could point this to any other dbt adapter.

Lastly, we do want to call out that the book is written to teach you what you need to know to get started being successful with dbt. We understand that there are many approaches in the dbt user community on how to use it and many that are heavily debated. Just go look at the dbt Slack channel, and you will inevitably run into some of these debates, especially with the topic of modeling. We approach the examples and instruction throughout the book based on our preference and experience, but we are in no way saying this is the only way. dbt is an extremely flexible product that can meet you where you are and support whatever decision you make. Additionally, there are some very advanced things that you can do with dbt, and for the most part, we won't cover these topics. We want this book to be accessible to even the most inexperienced dbt users, but maybe in the future there will be a book dedicated to advanced use cases!

Summary

In this chapter, we introduced you to what dbt is and the role that it plays in data analytics and warehouses. We learned that dbt is an open source command-line tool that helps Data Analysts and Data Engineers automate the process of transforming, modeling, and testing data in their organization's data warehouse. It was developed to make it easier to build and maintain analytics infrastructure, particularly in the context of data warehousing and business intelligence.

With dbt, you can write SQL and Python code using dbt models to define transformations and business logic that are applied to the data in your data warehouse. You can also use dbt to define and run tests to ensure the quality and consistency of your data. dbt is a fantastic tool due to its efficiency, collaborative nature, reliability, and the fact that it incorporates many software engineering best practices.

This chapter was intended to be a precursor to the rest of the book and create an overview of what to expect going forward. Don't worry if you are feeling overwhelmed as we are going to be spending a lot more time talking about each of the individual components outlined here.

CHAPTER 2

Setting Up a dbt Project

As with any new technical project, time must be spent on the setup and configuration to establish a solid foundation that will enable the contributors to the project to be successful. dbt is no different in this respect, and there is more than one path available to you for successfully setting up a new project.

As mentioned in Chapter 1, dbt is an open source project, which we will regularly refer to as dbt Core throughout the book. Given the nature of open source technology, you're able to install and configure your project in many ways. Throughout this chapter, we will make recommendations for installing dbt Core using pip, Homebrew, Docker, and cloning directly from GitHub. Additionally, we will address which installation option you should choose given the operating system you're using and your comfort level with the command line.

If installing and managing an open source project is not of interest to you or you have certain constraints that would limit you from directly installing dbt Core, there is an alternative solution that you can use called dbt Cloud. dbt Cloud is a software as a service tool that allows data professionals to run dbt jobs, develop within a browser-based IDE, and orchestrate your jobs. We will walk you through creating a dbt Cloud account and configuring a project within dbt Cloud.

Before diving into setting up a project in either dbt Core or dbt Cloud, we will compare the offerings of each solution and the factors that you should consider when choosing to implement dbt Core or dbt Cloud.

Tip As mentioned in Chapter 1, we will often reference the GitHub repository associated with this book. We recommend that you fork the repository if you are interested in following along with examples. It is hosted at `https://github.com/apress/unlocking-dbt`.

C. Cyr and D. Dorsey, *Unlocking dbt*, https://doi.org/10.1007/978-1-4842-9703-2_2

Comparing dbt Core and dbt Cloud

You have two options for how you can implement and manage your dbt project. You can install dbt Core for free and manage the infrastructure needed to run your project, or you can use dbt Cloud and all of the infrastructure, orchestration and scheduling, and documentation hosting will be managed for you. There are trade-offs associated with using either option including real costs, level of effort, complexity, and more. Throughout this section, we will provide a comparison of the fully open source solution, dbt Core, and the SaaS solution, dbt Cloud. You may be an experienced Analytics Engineer or someone completely new to the world of dbt who needs to stand up a dbt project from the ground up. This section can help you weigh the pros and cons to both dbt Core and dbt Cloud, so that you can make an informed decision on which solution you should implement for your use case. In Table 2-1, we have compiled some of the decision criteria that we will cover throughout this section.

Table 2-1. *Decision criteria*

	dbt Cloud	dbt Core
Vendor-managed infrastructure	✓	×
Free to use	×	✓
Customizable deployment	×	✓
CI/CD flexibility	✓	✓✓✓
Hosted documentation site	✓	×
Browser-based IDE	✓	×
Access to Cloud API	✓	×
Access to the semantic layer	✓	×
Out-of-the-box job scheduling	✓	×

The first option that we will cover is the open source project dbt Core. This is an appealing option to many because there is no direct cost for installing and using dbt Core. If you have productionized any projects or software using open source technology, you know that there are associated costs even though the open source technology is free to use. Additionally, using open source technology tends to be quick to set up locally, but productionizing your project will require a higher level of effort compared to a

SaaS product. Both of these statements hold when setting up and using dbt Core. That shouldn't deter you from wanting to use dbt Core, but it is important to be aware of these requirements.

If you are familiar with using the terminal, then it is very easy to set up dbt Core locally, and it is a popular option for engineers who want to explore the development process with dbt. You can install dbt Core via the terminal, Python, or Docker and build dbt models in a matter of minutes. We will discuss what the installation process looks like later in this chapter, but for now we will focus on when it makes sense for you to use dbt Core for your project. Where complexity comes in is once you have your dbt project to a state where you want to productionize it. Running dbt locally on your machine is not recommended for running dbt in production, so you will need to consider the architecture that will be required to run your project in production. To run dbt Core in production, you will need to set up infrastructure such as virtual machines, container registries, and/or services to run containers using a cloud provider such as AWS, Azure, or GCP. There are several paths forward for running your project in production, which we will discuss in Chapter 10. The main takeaway here is that for you to run a dbt Core project in production, you will need to set up, monitor, and maintain infrastructure. Additionally, you will need to consider the real costs associated with running dbt Core because the infrastructure requirements and the time spent by you come at a financial cost. These costs will vary depending on a multitude of factors, including the scale of infrastructure you implement and the complexity of your project.

Note Infrastructure isn't only an expense and maintenance burden. It also provides you the flexibility to decide the best approach for deploying dbt within your business. You have full control over the entire end-to-end process. For a team with DevOps or DataOps experience, this can be very lucrative. On the other hand, if you use dbt Cloud the infrastructure is completely hands off. Depending on the structure of your team, this may or may not be helpful.

Now that you are aware of the cost considerations and infrastructure requirements associated with building a project using dbt Core, let's consider when it makes sense to use dbt Core instead of dbt Cloud. If you are someone who is comfortable with software development workflows such as using git, using a code editor, and using the terminal and the effort and costs related to infrastructure fit within your constraints, then using dbt Core could be the right solution for you. Additionally, since dbt Core is an open source project, if you choose to implement it, you have a ton of flexibility in usage and customization. Lastly, the ability to maintain full control over the deployment process is one of the reasons that many people choose to implement dbt Core. The deployment process will look different across organizations, but common pieces include infrastructure, orchestration, and CI/CD. dbt Core provides flexibility in many ways, but not every team will have the resources available to maintain a dbt Core deployment.

So, when would you not want to use dbt Core? If you are not interested, or not comfortable, in maintaining the end-to-end deployment process associated with running dbt in production or don't have the bandwidth on your team, then dbt Core would not be an ideal solution for you. If you fall within this group, then consider dbt Cloud instead. Since dbt Cloud is a fully hosted browser-based solution, it takes a fraction of the time to build a new project from the ground up and deploy your project to production. There is no need to install dbt, manage and maintain infrastructure, or worry about orchestrating your dbt jobs. All of these features and more are built into dbt Cloud.

We will share how you set up dbt Cloud shortly, but first let's consider the offerings of this solution. First and foremost, dbt Cloud comes with a built-in integrated development environment (IDE) where you can develop right within your browser. Within the dbt IDE, there is a graphical file explorer that makes it very easy to traverse your project's directories. There is no need to use a terminal to change, move, or edit files. Depending on your skill set, you may like or dislike this. Additionally within the IDE, there is built-in git functionality, so you can create new branches, commit, pull, push, and start pull requests all from within the UI. Lastly, the IDE has a built-in pseudo-terminal that allows you to run dbt commands. This allows you to run and test models or build your entire project in your development environment all within the dbt IDE.

dbt Cloud has benefits beyond an intuitive IDE including job orchestration. Since dbt manages tracking the dependencies between all of your models, snapshots, tests, etc., it is very easy to orchestrate running your project in different environments in dbt Cloud. Within the Jobs section of the cloud UI, you have the ability to create jobs that point to different environments. Some common use cases include running a job in a production environment or running a CI/CD job to promote code to a production environment. Jobs in dbt Cloud can be triggered in various ways, such as selecting a basic hourly or daily interval, creating a custom cron schedule, webhooks triggered from

GitHub pull requests, events from an orchestration tool, or requests from the Cloud API. Some of these are more complex than others, but having the ability to scale up or scale down the complexity from the UI is a nice benefit.

There are additional perks to using dbt Cloud including hosted documentation and access to the dbt APIs. Regardless of whether you use dbt Cloud or dbt Core, you will always be able to generate documentation based on the YAML files within your project. The key difference is that when using dbt Cloud, you don't have to worry about where you host that documentation; it is simply available to users in the UI after generating docs. On the other hand, with dbt Core you have to find a solution for hosting your documentation. A common solution is to host it in an object storage platform such as AWS S3. We will discuss what this process looks like for dbt Core users in Chapter 10.

While there are many benefits to using dbt Cloud, it does come with trade-offs when you compare it to running dbt Core. dbt Core requires a higher level of effort to initially stand up and to maintain the long-term health of your project. By taking on that additional responsibility, you enable your team to have a greater level of flexibility with how to deploy dbt, how to implement CI/CD, change dbt Core code to fit your individual use case, and even contribute to the open source project. dbt Cloud is much faster at minimizing the time to value because you don't have to worry about managing deployments from end to end, but this of course comes with the trade-off of flexibility. As a dbt Cloud user, you are bound to the way things work within the UI, and if you have a need that is not being met, you will need to request a feature and wait for it to be implemented.

Now that we have discussed the trade-offs for implementing either solution, if you haven't already, you should have relevant knowledge to empower you to choose which solution best fits the requirements for your project. From here, we will move into how to set up and configure a dbt project using both dbt Core and dbt Cloud.

Note Throughout this book, we will use both dbt Cloud and dbt Core to provide you visual examples. Additionally, we will provide the code formatted directly as text within the book too.

Installing dbt Core

If you have decided to install dbt Core, this is the next stop in your journey to successfully setting up your project. In this section, we will cover the different ways that you can install dbt Core. There are several paths that you can follow for installation, so we will provide additional context regarding when it would be appropriate to use each method for installation.

Before we move into installing dbt, there are a few prerequisites that you should take care of before moving on. Since dbt Core is written in Python and distributed as a package, the first thing you should do is make sure that you have a recent version of Python installed. Before installing Python, check the dbt documentation to see what versions of Python are currently supported. Next, you should make sure that you have git installed. You can use whatever resource you are comfortable with to install git, but a common source would be to download and install git from git-scm.com, or if you're a Mac user, you can install git using Homebrew.

With these prerequisites in place, you are ready to start installing dbt. We will cover these four methods for installing dbt onto your machine:

1. Python pip
2. Homebrew
3. Docker
4. Cloning the source code from GitHub

We will first discuss installing dbt with pip. In our opinion, this is the best way to install dbt for local development. We will discuss other methods and their use cases later in this section.

Installing dbt with pip

If you have ever developed with Python, you know that it is very easy to end up in a dependency nightmare if you don't use virtual environments for local development. While this book isn't intended to teach you the best practices of developing locally with Python, we do recommend that you create a virtual environment for development before installing dbt. This will help you to maintain a tidy workspace for developing with dbt locally. For simplicity, we will use the *virtualenv* package to create a virtual environment, but in practice you may want to use a more modern package manager such as Pipenv or, our personal favorite, Poetry. To create a virtual environment, make sure that you have the *virtualenv* package installed. If you aren't sure if you have this package installed, you can run this command to both install and upgrade the package:

```
pip install virtualenv --upgrade
```

Now that you have the *virtualenv* package installed, you can create a virtual environment. We are going to create a virtual environment called *dbt-local-dev-env* that we will use throughout this book for development using dbt Core. To create your virtual environment, use the terminal to navigate to the directory where you plan on installing dbt; for us, that directory is named *my_first_dbt_project*. Once in the correct directory, you will run these commands to create your virtual environment and activate it:

```
python3 -m venv dbt-local-dev-env
source dbt-local-dev-env/bin/activate
```

Now that you have successfully created and activated a virtual environment, it's recommended that you install and upgrade both *wheel* and *setuptools*. This step is not required, but by having these two packages installed and up to date, your installation of dbt should be faster. If faster installation is appealing to you, you can run this command to get these packages installed and upgraded:

```
pip install pip wheel setuptools --upgrade
```

Once you have completed all of these steps, you are ready to install dbt using pip. There are two different ways that you can install dbt using pip. The first would be to install all of dbt Core, but this installs all of the adapters that dbt supports and is not recommended for most use cases. In the context of dbt, an adapter is how dbt connects with the data platform that you are using. There are several data platforms that dbt supports including Snowflake, Postgres, Redshift, BigQuery, etc. You can find a current list of verified and community-supported adapters here: `https://docs.getdbt.com/docs/supported-data-platforms`. Since we are going to be using Snowflake for the examples throughout this book, we will show you the installation process for the Snowflake adapter. To install the adapter of your choice, you will run the following command in the terminal, where "snowflake" should be replaced with the adapter that you wish to install:

```
pip install dbt-snowflake
```

You can check to see if the dbt-snowflake package was installed successfully using the command `pip show dbt-snowflake`. If the package was successfully installed and the output in your terminal looks similar to the output in Figure 2-1, then you now have dbt installed and you're ready to start developing!

```
(dbt-local-dev-env) my_first_dbt_project % pip show dbt-snowflake
Name: dbt-snowflake
Version: 1.5.0
Summary: The Snowflake adapter plugin for dbt
Home-page: https://github.com/dbt-labs/dbt-snowflake
Author: dbt Labs
Author-email: info@dbtlabs.com
License:
Location: /Users/ /my_first_dbt_project/dbt-local-dev-env/lib/python3.10/site-packages
Requires: dbt-core, snowflake-connector-python
Required-by:
```

Figure 2-1. *Output showing a successful installation of the Snowflake dbt adapter*

Tip Anytime you are going to stop working on your project, run the `deactivate` command in the terminal to deactivate your virtual environment. When you are ready to start working on your project again, run the `source dbt-local-dev-env/bin/activate` command to reactivate your virtual environment.

Installing dbt with Homebrew

Another method for installing dbt is using Homebrew, a popular package manager for MacOS. Homebrew makes it very easy to install dbt for local development, but there are a few limitations when using Homebrew to install dbt. You should only use this method to install dbt if you are going to be using MacOS for developing your dbt project and you plan to use one of these data platforms. Additionally, only use this method of installation if you don't care that dbt won't be installed in a virtual environment.

1. Snowflake
2. Redshift
3. BigQuery
4. Postgres

If you are going to be using MacOS for local development, and you don't plan on using one of these data platforms, then you should use one of the other methods discussed in this chapter to install dbt locally.

To get started with using Homebrew to install dbt, you should first make sure that you have Homebrew installed. You can do this by checking the version that you have installed by running this command in the terminal: `brew --version`. If a version is returned, then you are good to go; otherwise, you can visit `https://brew.sh` to get instructions for installing Homebrew. Once you have Homebrew installed, you will need to run the commands in Listing 2-1. These commands will ensure that brew is up to date, install git on your machine, and tap the dbt GitHub repository. The tap command doesn't actually install dbt, but instead it adds the dbt GitHub repository to the list that Homebrew will track for you. The last command instructs Homebrew to install the Snowflake dbt adapter. If you wish to install a different adapter, then replace `dbt-snowflake` with the adapter that you wish to install.

Listing 2-1. Commands needed to install dbt using Homebrew

```
brew update
brew install git
brew tap dbt-labs/dbt
brew install dbt-snowflake
```

As you can see, it is very straightforward to install dbt using Homebrew. You simply need to have Homebrew installed and run the four commands in Listing 2-1. Running these commands will install the latest version of dbt, but you can also use Homebrew to upgrade your dbt version when new releases are available. There are two ways that you can do this. The first would be to simply run `brew upgrade dbt-snowflake`, which will upgrade dbt to the latest stable version. The second and more advanced method would be to use Homebrew links which enable you to specify the version of dbt you would like to install and "link" that version so that when you run dbt commands the "linked" version is used. Imagine that when you ran the commands in Listing 2-1, version 1.2.0 of dbt Core was installed on your machine, but now version 1.3.0 is available as the latest stable version. You could use the commands in Listing 2-2 to unlink the current adapter, install dbt at version 1.3.0, and link v1.3.0 so that when you run dbt commands, v1.3.0 is used. The great thing about this is that Homebrew will retain v1.2.0, so you can toggle back and forth between the two versions as needed. As you can imagine, this provides a smooth way of upgrading to a new version of dbt without having to worry about breaking changes because you could always "link" back to the previous version.

Listing 2-2. Commands needed to install and upgrade dbt via brew link

```
brew unlink dbt-snowflake
brew install dbt-snowflake@1.3.0
brew link dbt-snowflake@1.3.0
```

Installing dbt Using Docker

While there are much simpler ways to install dbt on your local machine, using Docker to install and run dbt locally is an available option. Before we get into the details on how to use Docker to run dbt locally, let's first cover who should, and shouldn't, use this method for local development. First of all, Docker is an open source platform that allows you to automate the deployment, scaling, and management of applications using containerization. It provides an environment to package an application and its dependencies into a standardized unit called a container. Containers are isolated and lightweight, enabling them to run consistently across different computing environments, such as development machines, production servers, and cloud platforms. This essentially allows you to run your production deployment of dbt on your local machine.

We consider this an advanced way to run dbt, and if you are just starting out with dbt and you have never used Docker before, we do not recommend that you use the method to install and run dbt. However, if you are comfortable using Docker, this method can be advantageous because you will be running dbt locally similarly to how it will run in production. Additionally, you won't need to worry about setting up and managing a Python environment on your machine. If you are going to use this method for running dbt locally, make sure that you have a recent version of Docker installed.

To run dbt using Docker, you must have a dbt project and *profiles.yml* available for you to bind mount to the image. As mentioned in Chapter 1, we have set up a GitHub repository to go with this book that you can use for this. If you have cloned the book's GitHub repository, you will notice that we included a *profiles.yml* for you to help make it easy to set up your project.

Tip It is not good practice to check the *profiles.yml* file into version control if you store your sensitive information, such as account urls, usernames, and passwords, in it directly.

By default, dbt searches for this file in the present working directory, and if it doesn't find it, then it will check the *~/.dbt/* directory.

The *profiles.yml* in this book repository has been configured to work for Snowflake. If you will be using Snowflake, all you need to do is update the configurations to correspond to your Snowflake instance. You will need to update all of the configurations to correspond to your adapter's requirements. There are additional configurations that you can include in the file, such as a Snowflake query tag, connection retries, etc. We won't go into the details of the configurations required to set each of these up, but you can visit dbt's website, `https://docs.getdbt.com/reference/profiles.yml`, to view the additional profile configurations available for all adapters.

Tip Don't change the profile name for now, or you will run into issues with running dbt in Docker. We will cover additional profile configurations, including changing the profile name, in the next section of this chapter.

Once you have cloned the book's GitHub repository and set up a *profiles.yml*, you are ready to pull the dbt Docker image and start running dbt in Docker. To locate the image that you want to pull, you should visit the dbt-labs organization in GitHub here: `https://github.com/orgs/dbt-labs/packages`.

When you go to pull the Docker image, you can use tags to specify the version of dbt that you want to pull. Specifying a version is useful if you are wanting to upgrade or downgrade the version of dbt that you are using or have external dependencies that require a particular version of dbt. However, since we are building a new project, we will pull the latest version of dbt and tag it with a more meaningful name using these commands:

```
docker pull ghcr.io/dbt-labs/dbt-snowflake:latest
docker tag ghcr.io/dbt-labs/dbt-snowflake:latest dbt-snowflake-latest
```

With the image pulled and tagged, you can now run dbt locally using this image in a container. To do this, you will first want to make sure that in your terminal your present working directory is the directory where you cloned the book's repository. Once you have done that, you can run the command in Listing 2-3 to list out the files in the dbt project using the dbt command *ls*. Notice that in this command we are bind mounting two paths to the image. The first path is the current directory, represented by "`$(pwd)`". This is effectively mounting your dbt project to the image. The second path represents the directory where you placed your *profiles.yml*. If you followed along earlier in the chapter, then this should be the *~/.dbt/* directory.

Listing 2-3. Command used to run dbt in a Docker container

```
docker run \
--network=host \
--mount type=bind,source="$(pwd)",target=/usr/app \
--mount type=bind,source=/Users/username/.dbt,target=/root/.dbt/ \
dbt-snowflake-latest \
ls
```

Note Command-line examples throughout this book were done in bash, so you may need to change them if you are using a different terminal such as PowerShell.

Using the same command in Listing 2-3, you should be able to run other dbt commands using Docker. Try running the command again, but instead of the `ls` command, use the `seed` command. If you set up your profile correctly and dbt was able to connect to your data platform, Snowflake in our case, you should have seen a success message similar to the one in Figure 2-2. We will discuss what the `seed` command does in Chapter 3, but this is to help show you that you can run any dbt command in a container using this image.

```
00:51:56  1 of 4 START seed file public.raw_customers ................................... [RUN]
00:51:56  2 of 4 START seed file public.raw_orderitems .................................. [RUN]
00:51:56  3 of 4 START seed file public.raw_orders ...................................... [RUN]
00:51:56  4 of 4 START seed file public.raw_products .................................... [RUN]
00:51:59  4 of 4 OK loaded seed file public.raw_products ................................ [INSERT 10 in 3.12s]
00:51:59  3 of 4 OK loaded seed file public.raw_orders .................................. [INSERT 1000 in 3.21s]
00:51:59  1 of 4 OK loaded seed file public.raw_customers ............................... [INSERT 500 in 3.31s]
00:51:59  2 of 4 OK loaded seed file public.raw_orderitems .............................. [INSERT 2000 in 3.33s]
```

Figure 2-2. *Example of a success message resulting from running dbt seed in a Docker container*

Installing dbt from GitHub

As with the other installation methods that we have discussed, there are times when it does and doesn't make sense to install dbt directly from the source via cloning the GitHub repository. In our eyes, the main use case for installing dbt this way is if you are interested in contributing to dbt Core. After all, dbt wouldn't be what it is today without contributions from the community. Since you would be cloning the source code, this method provides you immense flexibility with the version of dbt that you are installing because you could clone the source code from a certain version or even an individual commit. With that said, most people don't need, or want, this level of flexibility when installing dbt for local development on their own project. For our purposes, we will clone the repository using the tag from the most recent stable release (at the time of writing this book).

Similarly to installing dbt with pip, before you get started with cloning the GitHub repository you should set up a Python virtual environment to keep everything tidy. For instructions on how to set up a virtual environment, see the section earlier in this chapter on installing dbt with pip.

Once you have a virtual environment setup, you are ready to clone the repository locally. For this step, you should keep in mind what adapter you want to use. If you are going to use Postgres, you can simply clone the dbt Core repository because this will include dbt-postgres by default. If you are going to be using Postgres, run the following commands from the directory where you would like to clone the repository to:

```
git clone -b v1.3.0 https://github.com/dbt-labs/dbt-core.git
cd dbt-core
pip install -r requirements.txt
```

Notice that we specified that we want to clone the repository based on the GitHub tag v1.3.0. By the time that you're reading this, dbt Core will likely have a new version, so you should adjust the tag to reflect that. Now, if you don't plan on using Postgres and would like to use a different adapter, such as Snowflake like us, you don't need to clone the dbt Core repository. Instead, you can clone the repository that corresponds to the adapter that you want to use. For example, if you wanted to use the Snowflake adapter, you would run the following commands to clone the repository and install dbt-snowflake:

```
git clone -b v1.3.0 https://github.com/dbt-labs/dbt-snowflake.git
cd dbt-core
pip install .
```

Once you have cloned the repository and installed dbt, you can validate that dbt was installed by running the command `dbt -version` in the terminal. This should return a result showing what version of dbt you have installed and if you are on the latest version. If you receive a message similar to this, then you have successfully installed dbt by cloning the source code from GitHub and are ready to move on to start initializing your dbt project.

Initializing a dbt Core Project

Up to this point, for dbt Core we have discussed the four different methods that you can use to install dbt. Installing dbt is just the first foundational step that you need to take before you can move on to start developing models with dbt. After installation, you need to initialize a project, but luckily dbt makes this very straightforward because of the `dbt init` command. This command will walk you through, step by step, initializing your dbt project. The command not only assists you with setting up a *profiles.yml*, but it will also build the file structure that dbt expects whenever you run other commands, such as `seed`, `run`, or `test`. Keep in mind that if you chose to run dbt in a Docker container, you will not need to initialize your project since you should already have a project directory and *profiles.yml* established. For all other installation methods that we discussed, you will need to go through this process to set up a new dbt project.

Note If you cloned the repository from GitHub for this book, we included a *profiles.yml* for you. Additionally, if you are using the repository for this book, you won't actually need to initialize a project, but this section will still serve as a good reference when you implement a greenfield dbt project.

To start initializing your project, you should make sure that your terminal's present working directory is where you want your project to be stored. Once you have navigated to the appropriate directory, you will run the `dbt init` CLI command. This will start the process of initializing your project, and a series of questions will appear in the terminal that you will need to answer. Most of the responses that you enter will end up being used as configurations in your *profiles.yml*. The questions that appear in the terminal will vary based on the adapter that you are using, so we will show you two different examples. The two adapters that we will use to initialize projects will be Snowflake and Redshift.

Tip Don't forget to activate your Python virtual environment where you installed dbt!

Let's start what this process looks like if you are using Snowflake. As mentioned earlier, you will want to run the `dbt init` command. For Snowflake, this includes naming your project, adding your Snowflake connection credentials, defining which database and schema dbt should build models in, and how many concurrent threads dbt should run. We chose to connect to Snowflake using a username and password. This is okay for us because we are developing locally with nonsensitive data. If you have heightened security requirements, consider connecting using a more secure authentication method such as a key pair or SSO.

Lastly, note that we have configured to run four concurrent threads. What this means is that when you execute a command such as `dbt run`, up to four concurrent queries will be sent to your database by dbt while conforming to the dependencies in your project's directed acyclic graph (DAG). While there is no maximum number of threads that you can set in your profile, it is recommended by dbt that you start with four threads and increase or decrease to optimize run times. Factors such as concurrency limits will impact the optimal number of threads for your project. Our best recommendation here is to experiment with the number of threads to see what works best for your project.

The series of questions to initialize a dbt project using the Redshift adapter is very similar to initializing a project using the Snowflake adapter. One notable difference is that you need to specify the port for connecting to Redshift. The default port, 5439, is listed, but be sure to enter the correct port for your Redshift cluster if it differs from the default port. Initializing your dbt project will vary slightly with other adapters because the configurations needed to connect to your data platform. With that said, regardless of which adapter you are using, the `dbt init` command provides an intuitive way to initialize your dbt project.

Now that you have the structure of your dbt project and *profiles.yml* in place, you should check to see if dbt can connect to your data platform that you configured during the initialization process. To do this, you can run the command `dbt debug`. This command will check three things: project configuration, required dependencies, and connection. With regard to project configuration, the debug command will check to see if your *profiles.yml* is in the expected directory and check to see if there is a *dbt_project.yml* file in your project directory. The connection test is one of the most important aspects of the debug command for a new project because this is how you validate if you configured your *profiles.yml* correctly. If the connection test doesn't pass, you should review your *profiles.yml* file and make sure that all of your credentials and connection information have been configured correctly.

If you open the profiles.yml that we included in the book's repo, you will notice that we have included three different connection configurations. The configurations are called **targets**, which is just a way to specify how to connect to your database and is how dbt Core is environmentally aware. We have included three targets: dev, prod, and ci. At the top of the file, notice that we have specified the default target to be dev. What this means is that when you invoke dbt if you don't specify a target, then it will use the dev target. For example:

```
dbt build (Will run against the dev target)
dbt build --target prod (Will run against the prod target)
```

This is the most common way to manage your environments in dbt Core. So, in a production job, you would want to add the `--target prod` flag to your commands so that database objects build in your production environment.

One final note is that you will most likely have many developers working in the same development database. So that developers don't overwrite each other's work, it is recommended that you uniquely name schemas per developer. A common pattern is to use your name as your developer schema. To do this, you simply update the schema config within the dev target of the *profiles.yml* file.

Tip YAML files are notorious for throwing errors due to inaccurate indentation. If everything is configured correctly in your *profiles.yml* and dbt is failing to connect to your database, then double-check your indentation! You may find benefit in finding a YAML validator online to help out with this.

Configuring a dbt Cloud Project

So far in this chapter, we have discussed dbt Core at length, and that's because there are several ways that you can install dbt locally. While running dbt locally provides you a great amount of flexibility, it does come with the trade-off of an increased level of effort needed to set up your project. In this section, we will discuss how dbt Cloud helps solve this problem by being fully browser based so you don't need to install anything locally and making it nearly effortless to set up a new project. We will show you the steps that you need to go through to set up dbt Cloud and give you a brief tour of the dbt Cloud IDE.

Plan Structure

Before we dive into setting up dbt Cloud, let's first cover the plan structure at a high level. Currently, there are three pricing tiers for dbt Cloud: Developer, Team, and Enterprise. Starting with the Developer plan will serve you well for following along with the exercises throughout this book. This plan is currently free for one user. Currently, this plan includes access to all of the main features of dbt Cloud, including the IDE, job scheduling, hosted documentation, and CI/CD.

Keep in mind when you create your account that if you want to get hands-on experience with the dbt APIs, you will need to do so during the limited time frame because this is only available on the Team plan and higher. That said, you could always upgrade to the Team plan. This plan is recommended for small teams that will have the need for up to two concurrently running jobs, want API access, and could use read-only seats to enable access to dbt-generated documentation.

The last is the Enterprise plan that comes with a whole host of features, including single sign-on, multiregion deployments, and role-based permissions. The Enterprise plan is recommended for large teams or organizations that have many distributed data teams. Pricing for the Enterprise plan is variable based on the size of your organization and requires you to work directly with dbt to come up with pricing. For the other plans, we intentionally have not included pricing as part of this overview of the dbt Cloud plan structure because it could change at any time. For the most current pricing and plan structure, please visit `www.getdbt.com/pricing`.

Getting Started

To get started setting up dbt Cloud, you will want to navigate to `www.getdbt.com/signup` and create your new account using the simple sign-up form. Once you sign up, you will need to verify your account creation via an email that will be sent to you by dbt Cloud.

Note We leave it up to you regarding where you decide to run the code examples included with the book's repo. You can use either dbt Cloud's free tier or you can run examples locally using dbt Core. However, the quickest way to get started learning is to use dbt Cloud because you don't need to worry about installing dbt.

Once you have verified your account, you will be able to start setting up a dbt project right from the browser. There are three main steps that you will need to take to get your project setup in dbt Cloud: name your project, configure an environment, and connect to a repository. If you read through how to install dbt Core locally, you likely notice that this process is similar to the `dbt init` command, but the process is more friendly to those who are not comfortable using the command line. Additionally, you should note that dbt Cloud is only available to users of a small subset of data platforms when compared to dbt Core. As referenced earlier in the chapter, dbt Core has over 30 available adapters, making dbt accessible to a very wide range of users. With that said, the dbt Cloud-supported platforms cover the vast majority of users who are implementing dbt, but if you aren't using one of these data platforms, consider seeing if there is a dbt Core adapter available for your use case. We'll refer you to dbt's documentation to get the most up-to-date information on which data platforms are supported by dbt Cloud.

To start setting up your project, you will first need to name your project. This can be anything that you would like it to be, but it should relate well to your core use case for dbt. Once you have your project named, you will need to select the data warehouse that you want to connect to and configure your development environment. The steps for connecting to your data warehouse will vary depending on which platform you are going to use. Let's walk through the configurations that are needed to connect to dbt to each of these platforms.

Note These configurations could change with future releases of dbt. To see the current setup instructions and required configurations, please visit `https://docs.getdbt.com/docs/get-started/connect-your-database`.

There are a few configurations that will be the same regardless of the database that you are connecting to. These configurations include the name of your development schema, the number of threads for your environment, and connecting to a repository. Unless you have a good reason to do so, we don't recommend that you change the name of your development schema from what dbt defaults it to. The default is the first letter of your first name and your last name and is prefixed with *dbt_*. With regard to setting the number of threads, please refer to the section on configuring a dbt Core project where we discussed how to decide the number of threads you should set for your environment. The last is connecting to a repository, which we will discuss further at the end of this section after covering the database-specific configurations.

Before we dive into platform-specific configurations, we wanted to call out that dbt does integrate directly with some platforms such as Snowflake and Databricks via a feature called Partner Connect. This enables you to simply connect dbt Cloud to your data warehouse platform instead of defining the connection criteria in the dbt Cloud UI. This greatly simplifies the setup by creating the needed accounts, permissions, and databases on your warehouse. It uses preset names for these that you can't change during the setup, so while it is really simple, it may not conform to your naming standards. If this is something you are interested in, you may want to check out the specifics from a combination of dbt Cloud's documentation and your data warehouse platform's documentation. For the rest of this section, we will talk about how to configure a connection directly in the dbt Cloud UI.

Connect to Snowflake

Connecting to Snowflake with dbt Cloud is a very simple process. When connecting to Snowflake, you will need to consider what authentication method you are going to use. For dbt Cloud users, there are three different methods you can use to authenticate the connection between dbt and Snowflake. First, you can connect by simply using your development username and password. This is the easiest way to get set up because you won't need to do any additional configuration in Snowflake if you already have a set of Snowflake credentials. If you don't want to enter your Snowflake password directly into dbt Cloud, you can use a key pair to connect instead. For this method, you will need to configure an encrypted key pair in Snowflake. The final method you can use for connecting to Snowflake is to use OAuth, but keep in mind that this authentication method is only available to dbt users on an Enterprise tier plan. Additionally, this authentication method, while the most secure, requires a fair amount of additional setup and requires periodic reauthentication. We will discuss each of these authentication methods later in this section.

Regardless of which authentication method you plan to use, you will need the following available for you to start configuring dbt to connect to Snowflake:

1. Snowflake account URL
2. Database
3. Warehouse
4. Role with read and write privileges to your target database

Once you have these items available, you can begin setting up a connection to Snowflake. When you are entering the account parameter into dbt Cloud, you will only need part of your account URL. For example, if your account URL is ab12345.us-east-2.aws.snowflakecomputing.com, you would only want to enter ab12345.us-east-2.aws. Next, you will need to enter a database, warehouse, and role in the corresponding fields in dbt Cloud.

Deviations start to arise based on the authentication method that you choose to use. The first is using a username and password. For this method, you simply need to enter your development username and password. Do note that at this point you should be setting up a development environment and should not use a username and password that correspond to a service account. We will address how to set up additional environments later in the book, and during this setup, it would be appropriate to use service account credentials.

While simply entering your username and password into dbt Cloud is the easiest way to connect to Snowflake, it may not be an option for you depending on the security requirements of your organization. In this case, we would recommend that you connect to Snowflake using key pair authentication. The first step you need to take is to create an encrypted key pair in your terminal by running this command to create a private key:

```
openssl genrsa 2048 | openssl pkcs8 -topk8 -v2 des3 -inform PEM -out
rsa_key.p8
```

When you run this command, we would recommend that you set a passphrase associated with the key. Be sure to save the encrypted private key and remember the passphrase because you will need to put this into the connection form in dbt Cloud. Next, you will need to create a public key and then set the public key for your user account in Snowflake. You can achieve this by running this command in the terminal to generate a public key:

```
openssl rsa -in rsa_key.p8 -pubout -out rsa_key.pub
```

Next, you will need to run this statement in Snowflake to assign this public key to your development user. This will need to be done by someone who has access to the `securityadmin`, or greater, role, so reach out to your Snowflake Admin if you need to. In this statement, you should replace *<your_username>* with the username that you will use for development and *<generated_public_key>* with the public key that you generated with the previous command:

```
alter user <your_username>
set rsa_public_key='<generated_public_key>';
```

After taking these steps, you should be able to complete the configuration in dbt Cloud. Figure 2-3 provides an example of what the configuration form should look like when using key pair authentication. Once your configuration is complete, you can test your connection to ensure that dbt is able to connect to Snowflake.

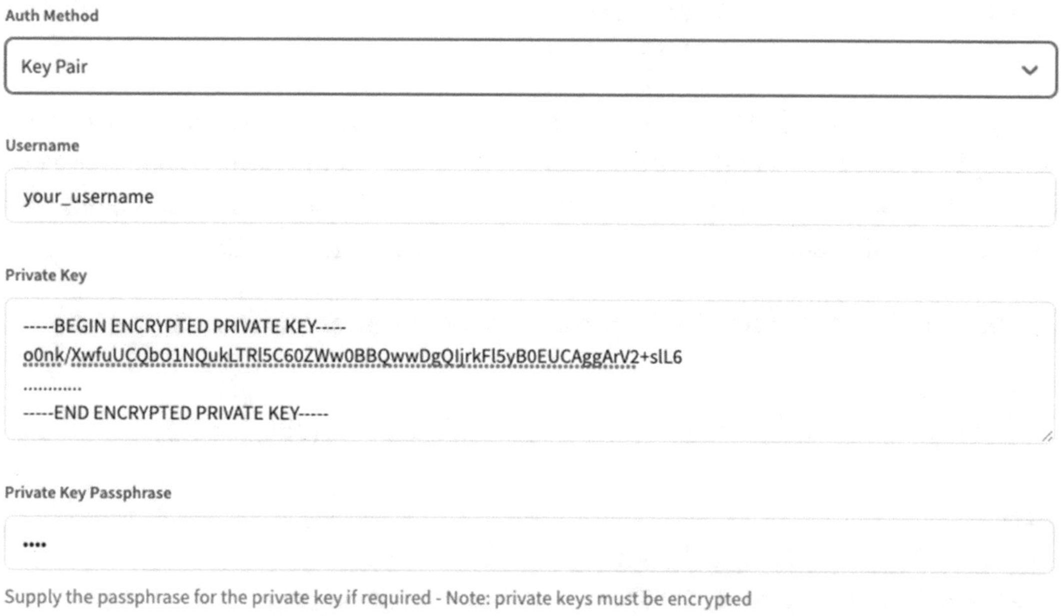

Figure 2-3. *Example of connecting dbt to Snowflake using a key pair*

Lastly, if you are signed up for the Enterprise tier, you have the opportunity to use OAuth to connect dbt to Snowflake. Keep in mind that this authentication method is not available in deployment environments. There are a few steps that you will need to take in Snowflake before you can go back to configure your dbt environment. If this is the authentication method that you plan to use, we recommend that you have your database administrator assist you because access to the `accountadmin` role in Snowflake or the `create integration` grant is required for this setup. The first step that you will need to take in Snowflake is to run the SQL in Listing 2-4 which will create a custom security integration named `dbt_cloud`. This command can be configured to fit your organization's needs, such as increasing or decreasing the time that the token is valid.

Listing 2-4. Command to create a security integration named dbt_cloud

```
create or replace security integration dbt_cloud
  type = oauth
  enabled = true
  oauth_client = custom
  oauth_client_type = 'confidential'
  oauth_redirect_uri = 'https://cloud.getdbt.com/complete/snowflake'
  oauth_issue_refresh_tokens = true
  oauth_refresh_token_validity = 86400;
```

Once you have created this security integration, you will need to retrieve the `oauth_client_id` and `oauth_client_secret` from dbt because you will have to enter these into dbt Cloud to complete the environment configuration. The dbt documentation provides the query in Listing 2-5 that can be used to retrieve these from Snowflake.

Listing 2-5. Query to retrieve the needed integration secrets from Snowflake

```
with integration_secrets as (
  select parse_json(system$show_oauth_client_secrets(dbt_cloud))
  as secrets)
select
  secrets:"OAUTH_CLIENT_ID"::string as client_id,
  secrets:"OAUTH_CLIENT_SECRET"::string as client_secret
from integration_secrets;
```

After retrieving these secrets from Snowflake using the query in Listing 2-5, you will need to enter these into dbt Cloud to complete the connection between dbt and Snowflake. As with all other authentication methods, you should test the connection in dbt Cloud before moving on.

Connect to Postgres or Redshift

For this section, we have grouped together Postgres and Redshift because the connection process is identical. When connecting dbt to Postgres or Redshift, you can configure it such that connection is made directly to your database, or you can elect to connect through an SSH tunnel. With regard to authentication, your only option for these databases is to use a username and password.

To set up the connection between dbt and one of these databases, you will need to have the hostname, port, and database that you wish to connect to. You will need these three values regardless of whether you are going to use an SSH tunnel or not. You can retrieve the hostname and port from the cloud platform you are using to host your database instance. The format of the hostname will vary slightly depending on which cloud platform you use, but the ports should be 5432 for Postgres and 5439 for Redshift unless you have explicitly changed these.

For additional security when connecting dbt to one of these databases, you can use an SSH tunnel to take advantage of port forwarding. We will assume that you already have an SSH bastion server set up. From this server, you will need to gather the hostname, port, and username and enter these values into dbt Cloud. Once you have saved this configuration in dbt Cloud, a public key will be generated for you that you will need to add to your SSH bastion server.

Connect to BigQuery

If you will be connecting dbt to BigQuery, there are two options for configuring your environment and authentication. The first option is to specify the connection configuration fields in dbt Cloud. You can either do this manually or create and upload a service account JSON keyfile. The latter is what we would recommend so that you can complete this process more quickly and accurately. To create a service account JSON keyfile, you should follow these steps:

1. Navigate to your Google Cloud console.
2. Select the project you will connect to dbt.
3. Select the email address you want to create a key for.
4. Select the keys tab, then create a new key.
5. Choose JSON as the file type and download.

Now that you have your keyfile available, you can upload this to the dbt Cloud environment configuration setup page. This will automatically fill in the values that dbt needs to connect to BigQuery.

If you are on an Enterprise tier plan, you can connect to BigQuery using OAuth to run queries in your development environment. Similar to Snowflake OAuth, this method cannot be used for authentication in deployment environments. Before you connect dbt to BigQuery using OAuth, there are a few prerequisites that will need to be completed in the BigQuery console by a user with administrator privileges. The first thing that you will need to do is navigate to the BigQuery console, select APIs & Services from the menu, and select credentials. On this page, you will be able to create a new OAuth client id and secret. When configuring this OAuth profile in the console, you will need to provide the following parameters:

1. Application Type: Web Application
2. Name: dbt Cloud
3. Authorized JavaScript Origins
4. Authorized Redirect URIs

Once you have configured this in the BigQuery console, you will need to copy the client id and secret because you will have to enter these into dbt Cloud. You will need to enter these into the corresponding client id and client secret fields in the connection settings in dbt Cloud. This will enable OAuth on your project, but you will still need to connect your developer environment to BigQuery. To do this, you will need to navigate to Profile Setting, then Credentials in dbt Cloud. From here, you should see an option that allows you to link your dbt Cloud account to your BigQuery account. Once you have successfully connected with single sign-on, your development environment is fully configured.

Connect to Databricks

In addition to Partner Connect, you could choose, or need, to connect to Databricks manually. The ODBC (Open Database Connectivity) connection to Databricks is flexible and allows you to connect to a cluster or SQL warehouse, but we will assume that you are already using Databricks and have one of these options set up.

Note If you need guidance on setting up a cluster or SQL warehouse, we refer you to the Databricks documentation. Set up a Databricks cluster: `https://docs.databricks.com/clusters/create-cluster.html`.

Set up a Databricks SQL warehouse: `https://docs.databricks.com/sql/admin/sql-endpoints.html#create-a-sql-warehouse`.

To get your project connected to Databricks, you will need to have a few configuration fields on hand, including

1. Hostname
2. Port
3. Cluster ID or SQL warehouse ID
4. Organization ID (optional)
5. User ID (optional)

Once you have these configurations available, you will want to navigate back to the dbt Cloud project setup form and enter these into their corresponding fields. Typically, the port will already be filled out for you in dbt Cloud with the default port of 443, but if you have configured Databricks to use a different port, you will need to update it. The setup form has fields available for both a cluster ID and a SQL warehouse ID, but you should only enter one of these depending on how you have configured Databricks. Figure 2-4 provides an example of what the connection parameters should look like in dbt Cloud for connecting to a Databricks SQL warehouse. Notice in the connection form that the SQL warehouse ID has been entered into the Endpoint field. This is because Databricks used to refer to SQL warehouses as SQL endpoints, but just know that the terms are synonymous. If you are connecting to a Databricks cluster, you should fill out the cluster field instead.

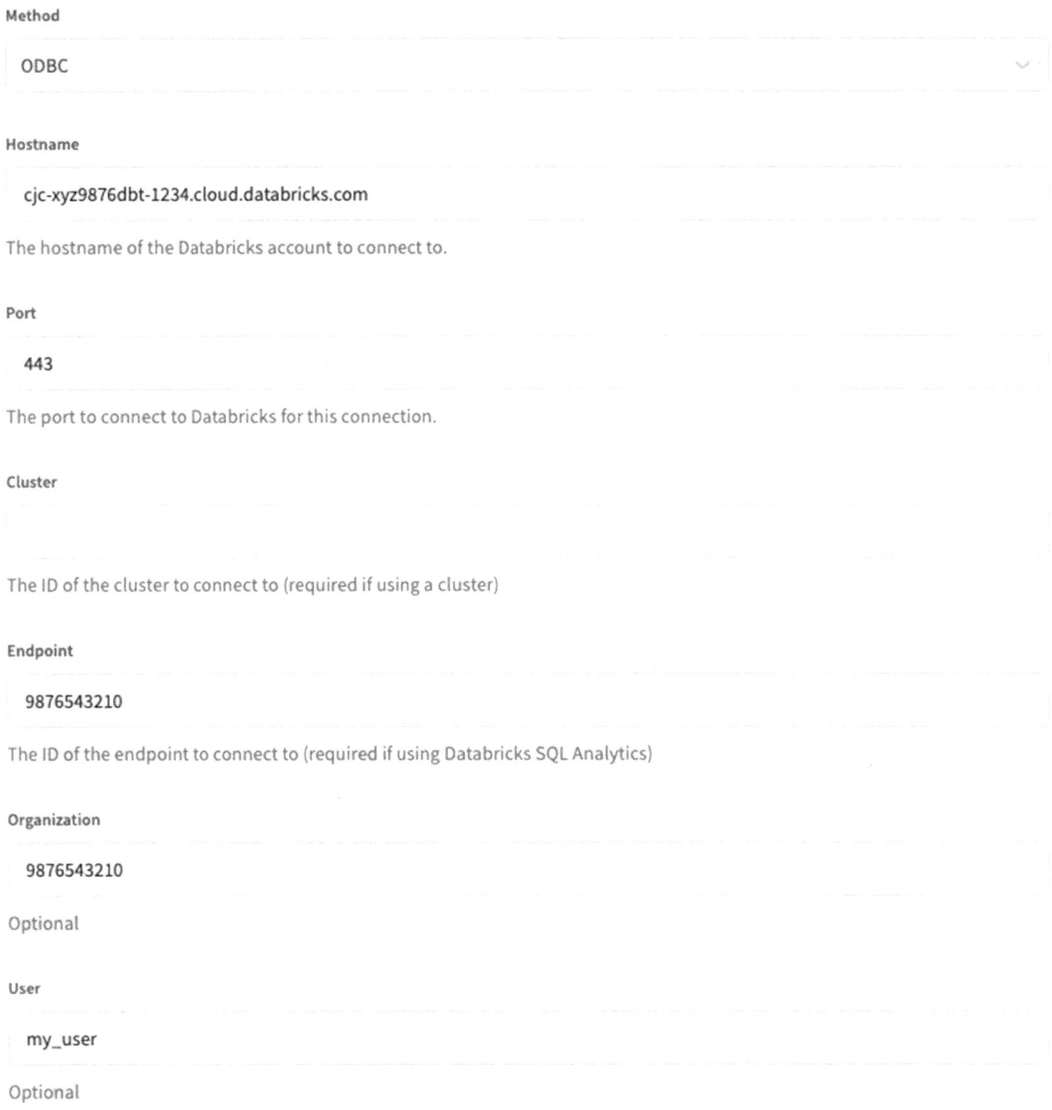

Method

ODBC

Hostname

cjc-xyz9876dbt-1234.cloud.databricks.com

The hostname of the Databricks account to connect to.

Port

443

The port to connect to Databricks for this connection.

Cluster

The ID of the cluster to connect to (required if using a cluster)

Endpoint

9876543210

The ID of the endpoint to connect to (required if using Databricks SQL Analytics)

Organization

9876543210

Optional

User

my_user

Optional

Figure 2-4. *Example configuration to connect to a Databricks SQL warehouse*

With the connection configured, the last step that you will need to take is to authorize your developer credentials. The only option for adding Databricks developer credentials to dbt Cloud is to use a personal access token generated from your Databricks workspace. If you don't already have a personal access token generated, you can generate it by navigating to the User Settings page in your Databricks workspace. From this page, click the Access Tokens tab and select Generate Token. You will need to copy this token into the developer credentials form in dbt Cloud. You should also store this token in a secure location in the instance that you need to use it in the future. Once you have completed these steps, you should be able to test your connection to Databricks. If your connection test is successful, you are ready to proceed.

Connect to Spark

The final data platform that we will cover connecting to dbt Cloud is Spark. When connecting to a Spark cluster, dbt Cloud supports connecting via HTTP or Thrift. The connection parameters in dbt Cloud for these methods are the same, but if you are interested in learning more about Thrift, we will refer you to the Apache Thrift documentation: `https://thrift.apache.org/`.

If you review the connection parameters needed to connect to Databricks, you will notice that they are very similar to those needed to connect to a Spark cluster. However, there are a few additional configurations that you can make when connecting to a Spark cluster. You will want to have these available before you continue with the setup form:

1. Hostname
2. Port
3. Cluster ID
4. Organization ID (optional)
5. User ID (optional)

Once you have these on hand, you will need to enter them into their appropriate fields in the project setup form in dbt Cloud. As mentioned, there are additional configurations that you can make when connecting to a Spark cluster. The first two additional configurations that you can make are connection timeout and connection retries. The setup form defaults these fields to a ten-second timeout and zero retries, but you can configure these to suit your needs. Lastly, if you are using the Kerberos network authentication protocol, you will need to supply values in the auth and kerberos service name fields. The auth field only accepts the value KERBEROS, which as you can imagine tells dbt that you want to connect using the Kerberos protocol. If you supply this value, you will also need to define the kerberos service name. Be sure to test your connection before moving on.

Set Up a Project Repository

One of the value propositions of dbt is to enable analytics and data teams to work more like software engineers. Of course, you aren't working like a software engineer if you aren't using git. Git is the de facto distributed version control system for developers, which is available for free use and download. While a full git tutorial is out of the scope for this book, there are a few concepts that you should be familiar with before moving on.

First, you should be familiar with branching; a branch is a version of the repository where you can freely change files without the worry of disrupting the current state of your repository. Next, you should understand that a commit is the act of saving your work to a branch. When you are ready to add work in your development branch, you will need to be familiar with merging and pull requests. Merging is exactly what it sounds like; you are merging one branch into another. Oftentimes, this includes merging your development branch into the main branch. Lastly, you should understand that a pull request is the request for a review of the code that you have changed prior to merging your work into a different branch. This is the bare minimum terminology you should be familiar with for working with git and dbt. However, we do recommend that you learn more about git, and a great reference is Scott Chacon and Ben Straub's book *Pro Git*.

With regard to setting up your project's repository, you have a few different options available to you when using dbt Cloud. The first option, and best for those who are testing out dbt Cloud, is to use a managed repository. To set up a dbt managed repository, all that you have to do is provide a name for the repository and click create. As mentioned, this method is ideal for those who are testing dbt Cloud during a proof of concept, but it is not advisable to use a managed repository if you plan to implement dbt in production. This is because when using a managed repository you have limited access to the full suite of git commands and features. Notably, you will not have access to pull requests, which is a critical part of git workflows.

For all git platforms, you can connect dbt Cloud to your repository by entering the repository's git URL into the setup form. For added security, dbt Cloud doesn't allow you to enter the https URL for the repository. Instead, you will need to provide either the *git@* or *ssh* URL. Once you have provided a valid URL to dbt Cloud, you will be provided with a deploy key. You will need to configure your git repository to provide read and write access using this deploy key.

If you are using a git platform that dbt supports a direct integration with, we recommend that you connect your repository this way. It is the easiest method for connecting an external git platform to dbt Cloud. Currently, there are three git platforms supported by dbt: GitHub, GitLab, and Azure DevOps. To create the connection between dbt Cloud and your git platform, you simply need press Connect in the setup form and follow the prompts to authorize dbt to connect to your git platform. Once you have authorized the connection, dbt Cloud will prompt you to select a repository to connect to.

You might ask, how do you connect to a repository that doesn't yet exist? Well, if you were building a project from scratch, we would recommend that you create an empty repository and then initialize the project in dbt Cloud. We will discuss how to initialize your project in dbt Cloud later in the chapter. However, for the purposes of this book, we have created a GitHub repository that you can fork and point dbt Cloud to. If you haven't already forked this repository, see the introduction to this chapter for the link to the repository.

Caution The dbt Cloud connection with Azure DevOps is only available to users on the Enterprise tier plan at the time of this writing.

Regardless of which method you used to connect to your project's git repository, you have successfully configured your dbt Cloud project. That means that you are ready to move on to initializing your project and getting a scalable project structure set up.

Initializing a dbt Cloud Project

Now that you have successfully configured your dbt Cloud project, you will need to initialize your project. Within the scope of dbt Cloud, initializing your project means that dbt will automatically build out the basic expected file structure in your git repository. This includes added directories for models, seeds, and more, but more importantly it adds a *dbt_project.yml* file. This file is used to configure many different aspects of your project, such as materializations, database and schema overrides, and hooks.

To initialize your dbt project, you will need to navigate to the development IDE within dbt Cloud. To get there, you will need to select the Develop tab from the home page. Once the Cloud IDE loads, you should click the button that shows "Initialize dbt project" in the top-left corner, similar to the one seen in Figure 2-5. Clicking this button will add several files and directories to your dbt project, but keep in mind that these have only been added locally. To actually get these changes added to your git repository, you will want to click the git-control button again. This time, the button should show "Commit and Push." Clicking this will add a commit to your main branch and push the changes to your remote repository, but first a window will pop up allowing you to add a commit message. You will want to add a description such as "Initialized dbt project structure." As you can see, the dbt Cloud IDE allows you to easily use basic features of git, such as commit, push, pull, etc., all from this one button. If you click the drop-down menu in the git-control button, you can also create branches and revert commits all with the click of a button as well.

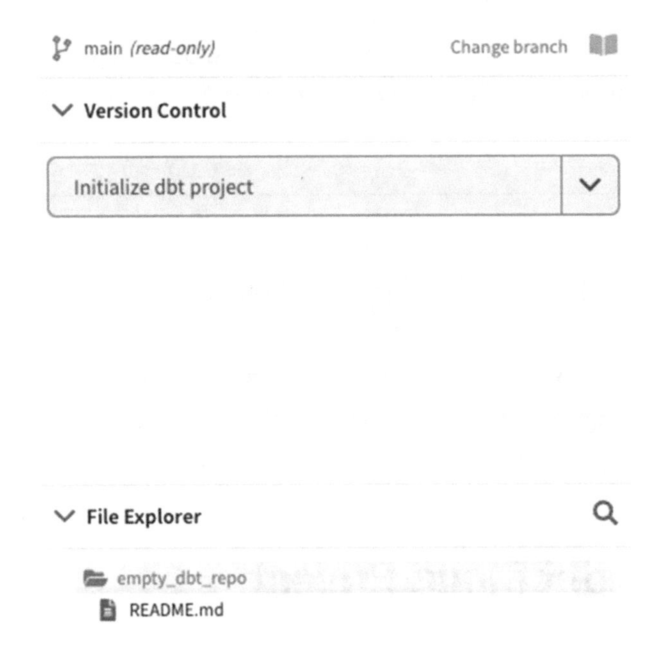

Figure 2-5. *Git-control button in the dbt Cloud IDE ready to initialize a project*

Project Structure

No matter whether you are using dbt Cloud or you are developing locally with dbt Core, when you initialize your project the file and directory structure that dbt builds for you is the same. In this section, we will discuss ways that you can organize the structure of your project. It is important that you establish a standardized project structure early on, ideally before any development, so that you and your team can focus on the important things and not be worrying about where they should store a new model.

When discussing project structure, we will make recommendations, but whether or not you choose to use our recommendation or you create your own structure doesn't really matter. The most important thing is that your structure becomes the standard for your project and is communicated to your team. Since dbt is meant to be used by a team of developers and analysts to collaboratively build, if you don't have a standardized structure your project directories will quickly spiral into a mess. First, we will discuss dbt's recommended project structure, but that will be followed by a few enhancements to this base structure that we have seen work well in projects that we have worked on.

Let's start at the top level of the project before we get more granular and discuss the fine details of structure. As discussed in Chapter 1, there are several directories that are part of every dbt project, including models, snapshots, seed, macros, etc. These directories are where dbt searches for different asset types at compile time. For example, when you execute the `dbt run` command, dbt will search the models directory and execute every defined model. No matter how you decide to structure your project, it is important to understand that you shouldn't change the name of these directories unless you have a good reason to do so.

Tip While we caution you to not change the top-level directory names, that doesn't mean you can't. If you have a good reason to do so, you can update the *dbt_project.yml* to search for your renamed directories.

In addition to the top-level directories, there are a few files that should live at the root directory of your project. At minimum, you should have these files: .gitignore, dbt_project.yml, packages.yml, and README.md. While you don't need a packages.yml or README.md for your project to run, we do recommend that you include these files when you create your project. The packages.yml will be used to take advantage of using packages that have been generated by the dbt community within your project. Additionally, every well-structured project repository should include a README.md so that anyone can quickly gain an understanding of the purpose of your dbt project. The main directory that you should be concerned with when structuring your project is the models directory, which we will discuss in the next section.

Models Directory Structure

Since dbt is all about implementing data transformation and making them available to end users, it should be clear the models are the core of any dbt project. As such, let's first discuss recommendations for structuring the Models directory. But, before we move on, we want to call out that this directory structure heavily leans toward building a dimensional model (star schema). However, we want to acknowledge that this is not the only way to structure your project nor is it the only way to model your data. That is to say, throughout this book we will be following this structure because it is a common pattern that we see many teams follow that use dbt. There are numerous ways that you could structure your project, and as mentioned before, the most important thing is to have a plan and stick to it.

A common pattern for thinking about how models should be organized and structured can fall into one of three categories: staging, intermediate, and marts. In this pattern, every model you develop should fit into one of these three categories. In Chapter 4, we will discuss in detail what these different levels of models are and how you should implement and use them. But, at a high level, staging models are a cleaned version of your source data, intermediate models store transformations that are used downstream multiple times, and mart models are what you will expose to your end users.

However, much like everything else, this is not the only way to do things, but it is a common pattern. Another common pattern is to use staging models as a place to conduct your final transformations, run tests on them, and then copy them over to mart models. This pattern is often referred to as the Write-Audit-Publish (WAP) pattern. It is a way for you to validate that data is accurate before it ever gets written to production database objects. We will leave the final decision up to you as to how you should structure your project.

Figure 2-6 provides a visual example of how the models directory might look if you plan to structure your project following this recommendation. First, notice that under the models directory sit three subdirectories that correspond with the three levels of models.

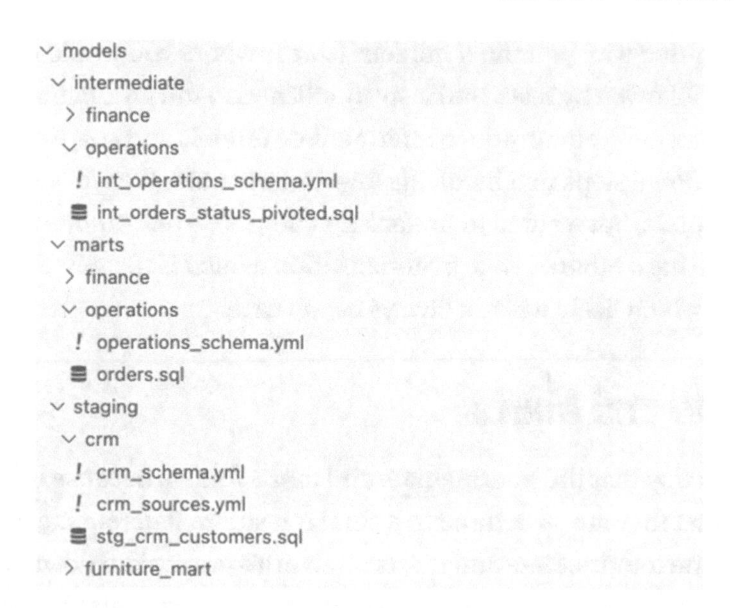

Figure 2-6. *Recommended models directory structure*

Staging

Let's break down these three subdirectories and how we recommend that you structure them starting with the staging directory. The staging directory is one that will likely change based on how you plan to structure your transformations. If you plan to keep staging models one-to-one with your source data, then you should create a folder for each of your source systems that you receive data from. In the case of our example project, data is received from a customer relationship management system (CRM) and from a product database identified as Furniture Mart. Having one folder per source keeps your project structure very clear as to what system the staging model corresponds to. For example, take the *stg_crm_customers* model in Figure 2-6. The CRM is where customer data is coming from for this project, so we nest that model in the crm directory. Also, notice how the model is named and that there are three parts it can be broken down into. Stg identifies that this is a staging model, crm identifies the source system, and customers is a description of what the model actually is. If you are taking this approach to staging models, then we recommend that you follow this naming convention for all staged source models.

However, if you plan to follow a different pattern, you will probably change how you think about staging models. For example, if you are following the Write-Audit-Publish pattern, instead of your staging models being one-to-one with sources they would instead be one-to-one with your final models. Here is where you would conduct your transformations, followed by tests, and lastly published to the production assets. One caveat here is that when you introduce intermediate models and snapshots in dbt, these may require additional steps that break the one-to-one mold. We will cover those more in upcoming chapters, but wanted to preface that here. The one-to-one comments here assume you are using a table or view materialization, which is recreated every time you run your models which likely will not always be the case.

Intermediate and Marts

We have grouped together the intermediate and marts folders because it is recommended that they are structured in a similar fashion. Referencing back to Figure 2-6, both the intermediate and marts directories are broken down into different departments. If you are familiar with the concept of data marts, this should feel familiar, but if you aren't, you should know that a data mart is a subset of the data warehouse that has been curated specifically for a business unit, subject, or department. They allow you to tailor your data model in a way so that the intent is clear and provides teams to quickly obtain relevant information for decision-making and analysis. In our example, we have generated a finance mart and operations mart. This structure allows you to build data mart–specific intermediate and user-facing transformations.

Similar to the staging directory, you will see a schema YAML file for both the marts and intermediate directories. We covered this lightly in Chapter 1 and will dive more into those in the next chapter. These should also be used to define the column names, descriptions, and tests for models in these folders. Additionally, we recommend that you prefix your intermediate models with int_.

Masterdata

So far, we have the project structure that we have recommended aligns very similarly to the project structure that dbt recommends as well. However, there is one deviation that we like to make when building a dbt project which is the addition of a masterdata folder. Let's go back to the brief description of a data mart and expand on it slightly more.

Each of your data marts is typically made up of two types of models: facts and dimensions. These concepts come from Ralph Kimball's dimension model. Facts are tables that measure events such as orders in this book's example project. In fact (no pun intended), data marts are simply just a collection of fact tables, and it is very clear which mart they belong to. For example, a revenue model would fit into the finance data mart very clearly.

On the other hand, we have dimension models which are datasets that are used in combination with facts to describe, group, and filter the events that are being measured. An example of a dimension would be customers. If you are trying to create a centralized view of a customer that you will reference in more than one data mart, it is difficult to figure out where to store this model if you follow the base structure recommendation. This is where the masterdata directory comes in handy. Masterdata is a concept that has been around for as long as analytics has happened, and numerous tools have been created to help companies with this. It creates a single view of unique information that describes a business's core entities. In the example of a customer, you may have multiple source systems that contain information about a customer. For example, you may have a product database, a CRM, a financial system, etc. that all have their unique ways of describing a customer. The masterdata concept takes all of this information, pulls it together, to create a singular way to look at the customer. A customer is just one example, but you likely have several examples of this within your organization of ways that you could unify data definitions via master datasets.

We recommend that you use this directory to store dimension tables that you will reference in more than one data mart. Figure 2-7 shows how we have changed the project structure slightly to include a masterdata directory with a customers model that will be exposed to end users.

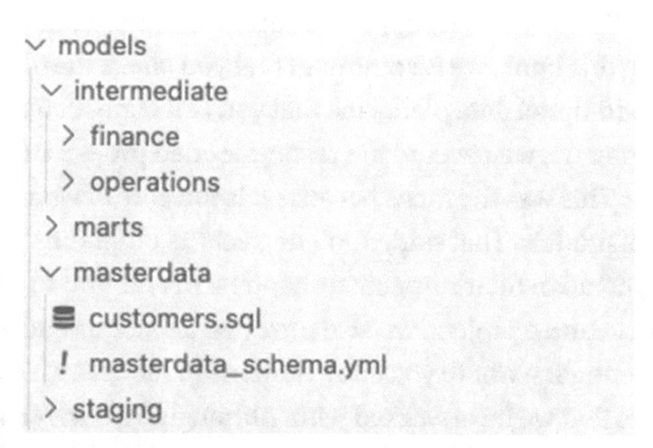

Figure 2-7. *Masterdata folder to store dimensions*

Summary

As with most projects you work on, it is typically worthwhile to take the time to set everything up correctly from the beginning. Considering you are reading this book, you have taken the first steps to get everything right from the beginning. One of the most important decisions that you have to make when considering an implementation of dbt is to decide if you are going to implement dbt Core or dbt Cloud. This chapter discussed the pros, cons, and trade-offs of implementing both solutions. The beginning of this chapter should have left you feeling empowered with the required knowledge to make this important decision.

Although extremely important, choosing to implement dbt Core or dbt Cloud is just the first step to success. Once you have this decision made, you will need to go through the process of setting up a project. This process varies depending on whether you are using dbt Core or dbt Cloud. For example, with dbt Cloud there is nothing you have to install, but the opposite is true for dbt Core. With dbt Core, not only do you need to install dbt, but you have to decide which method you are going to use for installation. We covered the four methods that can be used to install dbt Core and when each installation method is appropriate.

On the other hand, for those that have selected to implement dbt Cloud, the setup process is significantly easier. We shared how you can set up a project using different data platforms, such as Snowflake, BigQuery, and Databricks. While we discussed the setup process for every platform that dbt Cloud is compatible with, there may be additional platforms that dbt Cloud supports in the future. Depending on how far in the future you're reading this book, we recommend that you check the dbt documentation to see if there are any additional data platforms that you can connect to.

To wrap the chapter up, we covered the recommended project structure with a focus on the models folder. This was the focus because it is the core of what you will work with when developing dbt models. That said, don't neglect the other directories, and we will cover them in more detail in their respective chapters. In addition to covering how dbt recommends you structure a project, we also provided a recommended enhancement to this structure to centralize where you store dimension models. This has proven to be successful in projects that we have worked with. Although we covered a recommended project structure, you shouldn't let this structure limit what you do with dbt. If you have the need for additional folder structure, that is absolutely fine. The most important thing is that you define a standard for your project and stick to it. In the upcoming chapter, we are going to cover data sources and seed files. These concepts within dbt will continue setting you up to start developing data models.

A final note on project structure is that you can do whatever you think will work best for you and the data model that you plan to implement. In our examples, we favored a dimensional model, but don't think that this is the only way to do things! In fact, there are entire packages that help you set up projects for other types of data models. For example, AutomateDV is a package designed entirely around using dbt to implement a Data Vault model, and as you can imagine, if you followed our example project structure you would utterly fail at implementing a Data Vault. We want to keep our examples as agnostic as possible, but we could end up writing an entire book on the numerous different ways to structure your project (but we imagine that isn't why you're reading this!).

References

dbt Documentation: `https://docs.getdbt.com/docs/introduction`
Python Virtual Environment: `https://docs.python.org/3/library/venv.html`
Homebrew: `https://brew.sh/`
Docker: `https://docs.docker.com/`
Snowflake Key Pair: `https://docs.snowflake.com/en/user-guide/key-pair-auth.html`
BigQuery Service Account Key File: `https://cloud.google.com/bigquery/docs/authentication/service-account-file`
Databricks Personal Access Token: `https://docs.databricks.com/dev-tools/api/latest/authentication.html`
Git: `https://git-scm.com/`
Pro Git, Second Edition (2014), by Scott Chacon and Ben Straub

CHAPTER 3

Sources and Seeds

In this chapter, we will explore the concept of sources and seeds in dbt. These two items are mostly independent of each other, and we combined them into the same chapter as they are smaller, but very important, subjects. Sources refer to the original data sources, such as data from databases, that are used as the input for dbt. Source tables are referred to in dbt models via a Jinja function, which we will discuss more later in the chapter. Seeds, on the other hand, refer to the data that is used to populate the initial sets on small unchanging data, typically used for reference data or lookup tables. Seeds are still referenced within models utilizing a Jinja function, but use a different command than sources.

In this chapter, we will discuss best practices for working with sources and seeds in dbt and provide examples of how to use them in real-world scenarios. We will look at what they are, the benefits of using them, and common ways to use and configure them. It is helpful to have a basic understanding of models (covered in Chapter 4), but we believe that it has been introduced enough in Chapter 1 to be able to make sense of it here. We will cover this more in depth in the next chapter though.

What Are Sources?

In dbt, a *source* references databases, schemas, and tables in your connected database from which dbt can read data. They are defined within YAML files and referenced throughout your model creation. There are several reasons why you will want to make this a standard practice when setting up your dbt project. Instead of having to reference a data location in every script or model, sources allow you to name and define it once using your *sources.yml* file. This is especially important when you start working with more than one environment (i.e., development, QA, production) that live in different databases or schemas. We will dive more into this benefit and several others in the next several sections.

C. Cyr and D. Dorsey, *Unlocking dbt*, https://doi.org/10.1007/978-1-4842-9703-2_3

Note It is very common to have multiple sources.yml files, especially when you have a larger dbt project. You cannot use the same name twice and will need to name them something that is identifiable to the developer. Typically, each source YAML file supports a single folder or group of folders in a model hierarchy and is named something that references their connection.

By default, dbt does not create a source file for you so will need to create it yourself. There are many different tactics on how to do this, but getting started we recommend creating your sources.yml files in the root of the models folder and storing everything there. You can name the files whatever you like, but for simplicity we recommend just starting with *sources.yml.* Also, as your sources grow, you may want to consider adding additional source files and that is fine. dbt provides you with the flexibility to set it up however you like. When you first create your sources.yml file, it will obviously be blank. So to make use of it, you will need to add the correct YAML syntax. YAML syntax is very specific about spacing and tabs, so don't worry if you mess up a few times getting started. You will get the hang of it. If you are using dbt Cloud, then dbt is really good about identifying the mistake as well and letting you know what to fix.

Reminder You can only create sources for data that already lives in your dbt target data store.

Sources are defined in your source YAML files nested under the *sources:* key. You can set your sources.yml file to capture basic information, or there are several other things you can add to expand on it. For simplicity sake, let's start with just looking at the basics. Let's assume we have a database called raw that exists in the data store that we are connecting dbt to. The raw database has multiple schemas and tables associated with those tables. In our example, we have a schema called product and two tables called items and purchases. At a basic level, Figure 3-1 shows how we would structure this in our YAML file.

```yaml
version: 2

sources:
  - name: Product
    database: raw
    schema: product
    tables:
      - name: items
      - name: purchases
```

Figure 3-1. *Basic sources.yml file*

The name is the name of the source and the first value you will use in the source Jinja function we will show later. The naming is pretty intuitive, but nonetheless we want to be explicit in describing what each config means. The database is the name of the database. The schema is the name of the schema in your database. And finally, the tables are the name of your tables. The schema name is the only item not required in the example, but we think it is best practice to always include it as if it is. If you do not use the schema value, then the name value (the source name) is interpreted as the schema name.

You may also be wondering about the "version: 2" and what that means. In the early days of dbt, the way that YAML files looked was very different to the version we see now, so they added the version to allow them to make the structure more extensible. Even though version 2 is currently the only supported version of these files, they kept it around as a required key in case they introduce a new file structure down the road. It will be a lot easier having that there than the alternatives. Despite it currently being static, this value will need to always exist in your sources.yml file for it to work. Additionally, it will need to exist in any YAML files that you want dbt to use within your project.

Now that we have looked at some of the basic components of your source YAML file, let's take a look at other things you can also do with it. Let's start by taking the basic example shown in Figure 3-1 and add some additional items shown in Figure 3-2.

```
1   version: 2
2
3   sources:
4     - name: Product
5       description: This is a copy of my Product database
6       database: Product
7       schema: public
8       tables:
9         - name: items
10          freshness:
11            warn_after: {count: 24, period: hour}
12            error_after: {count: 48, period: hour}
13          loaded_at_field: Modified_At
14          columns:
15              - name: id
16                description: Primary key of the items table
17                tests:
18                  - not_null
19                  - unique
20        - name: purchases
```

Figure 3-2. *Basic sources.yml file with additional values*

Here, you can see that we added values for description, source freshness, and tests. Descriptions can be added to your YAML files to document objects in it. These are not exclusive to your sources.yml and can also be used in your schema.yml files that will be covered in Chapter 4. These are very valuable because they are used when you build out your documentation site. dbt reads from your YAML files and populates the information for you. We have dedicated Chapter 9 of the book to documentation and will cover this topic more there.

We also have the ability to add tests within this file. As you can see in Figure 3-2, we have used two built-in dbt tests (not_null and unique) on the "id" column of the "items" table. These tests make sure that the "id" column does not have any NULL values and that the "id" only contains unique values before this source gets processed in downstream models. We have not covered tests in the book yet, but we will cover it in greater detail in Chapter 8. However, we wanted to go ahead and point out that it is possible to use tests on models and sources.

The last new option you see is freshness. This is an option that checks whether you have stale data that hasn't been updated in a configurable amount of time. It will either trigger a warning or an error. We will cover this more in a later section of this chapter.

Once you have defined your sources in sources.yml, you can use them in your dbt models and tests by referencing the source name. To do this, you will utilize the Jinja function named source which looks like this: *{{ source('source_name', 'table_name') }}*. This will allow dbt to read data from the *'table_name'* table in the *'source_name'* source when running the model. More on this later in the chapter.

Complete Source Properties Syntax

We have touched on some of the options of what you can do with your sources.yml file, but we wanted to provide you with the full YAML syntax. The following highlights all of the currently supported options you can add to your sources.yml file:

```
version: 2
sources:
  - name: <string> # required
    description: <markdown_string>
    database: <database_name>
    schema: <schema_name>
    loader: <string>
    loaded_at_field: <column_name>
    meta: {<dictionary>}
   tags: [<string>]

    overrides: <string>

    freshness:
      warn_after:
        count: <positive_integer>
        period: : minute | hour | day
      error_after:
        count: <positive_integer>
        period: minute | hour | day
      filter: <where-condition>

    quoting:
      database: true | false
      schema: true | false
      identifier: true | false
```

```
  tables:
    - name: <string> #required
      description: <markdown_string>
      meta: {<dictionary>}
      identifier: <table_name>
      loaded_at_field: <column_name>
      tests:
        - <test>
        - ... # declare additional tests
      tags: [<string>]
       freshness:
        warn_after:
          count: <positive_integer>
          period: : minute | hour | day
        error_after:
          count: <positive_integer>
          period: minute | hour | day
        filter: <where-condition>

      quoting:
        database: true | false
        schema: true | false
        identifier: true | false
      external: {<dictionary>}
      columns:
        - name: <column_name> # required
          description: <markdown_string>
          meta: {<dictionary>}
          quote: true | false
          tests:
            - <test>
            - ... # declare additional tests
          tags: [<string>]
        - name: ... # declare properties of additional columns
    - name: ... # declare properties of additional source tables
- name: ... # declare properties of additional sources
```

As you can see, there are a considerable number of options for how to configure your sources.yml file. We will continue to hit on several of these throughout the rest of the chapter and the book.

Benefits of Sources

We have already touched on several benefits of using sources in your dbt project, but let's dig into more of the specifics:

Reusability: By defining your sources in a central location (the sources.yml file), you can reuse them across multiple models in your dbt project. You don't need to hardcode the source information into every model that you build that references it.

Maintainability: Defining your sources in sources.yml allows you to separate your data source locations (database, schema, and table names) from your actual dbt code. This makes it easier to maintain your dbt project, as you can update your source location details in one place rather than having to update them in multiple models and tests. You can also reference environment variables in your source files if you are pushing code between multiple environments (development, QA, production, etc.) and your schema objects are labeled differently. We will cover this in more detail later in the book.

Consistency: By using sources, you are storing all connection-related details in one place. If a database or schema name changes, you don't have to update your code in every model that references it, you only have to update it in one place. If you have multiple developers working in the same project, then everyone knows where the connections are referenced. Additionally, you have one place to add descriptions and tests to your source data.

Lineage: Utilizing sources alongside the source() Jinja function will affect model run order (dbt DAGs). If there are any tests, source freshness checks, etc. that need to happen first, then this ensures they do happen first.

We believe that creating and managing sources in your dbt project is a core competency in dbt and something you should absolutely do.

Referencing Sources in Models

The primary use case for using sources is to reference them in your transformation models. To do this, you need to utilize the source Jinja function. The source Jinja function replaces the parts of a SQL statement that you use for table names with the following syntax: *{{ source('source name','table name') }}*. In the syntax, you replace the *source name* with the name of your source and the *table name* with the name of your table. Using what we showed in Figure 3-2, let's look at an example in Listing 3-1.

Listing 3-1. Basic query using the source function

```
select
  items.*
from {{ source('Product', 'items') }} as items
inner join {{ source('Product', 'purchases') }} as purchases
  on items.items_id = items.id
```

In the example, we have entered the name of our source for both tables (Product) and the name of the table we are reading from (items and purchases). If we look at how dbt compiles the query, it will look like what is shown in Listing 3-2.

Listing 3-2. Compiled query of Listing 3-1

```
select
  items.*
from Product.public.items as items
inner join Product.public.purchases as purchases
  on items.items_id = items.id
```

Pro Tip When you are developing in dbt Cloud, you can enter __ (two underscores) to populate the source syntax and syntax for many other Jinja functions.

You can only reference sources that exist in your sources.yml file. If they do not exist, then you will get an error until you add it. If the source exists and you are still getting an error, then check to make sure your casing is the same. Many parts of dbt are case sensitive, and this is one of them. As authors of this book, when we first started learning dbt we spent a lot of time troubleshooting problems that ended up being as simple as having something capitalized that shouldn't have been. It is also worth mentioning that spaces and proper quoting are important and can also lead to failures.

Note You can only reference sources that exist in your sources.yml file. If they do not exist, then you will get an error until you add it.

Other problems we commonly see are mistakes made in the sources.yml file or in adding the source to your models. Here are some things to check if things aren't working:

- Make sure you populated your schema.yml file with the values in the right place. For example, make sure that you didn't inadvertently put a database name where the schema name should go.
- Check to make sure that the source you are referencing actually exists in your data store. This may feel obvious, but sometimes when you are cranking through populating your sources, you may be relying on what you think instead of what you know.
- Make sure the account you are using to connect dbt to the data store has access to the source destination.
- Make sure your letter casing is the same in your source Jinja function as it is in your sources.yml file. If in one place you are capitalizing the first letter and in another making it lower case, then it could be a problem.

Source Freshness

As we have discussed, dbt is only a transformation tool and thus relies on other data pipelines, services, and/or processes to land the data in your data store first. Having data that has already been ingested to your data store is a critical aspect of dbt functioning correctly even though it is managed separately. Thankfully, dbt provides us a way to help with that using source freshness. With just a couple of extra configurations, dbt can monitor the "freshness" of the data in your source tables. This is very useful in making sure your upstream data pipelines are in a healthy state and you still have new data coming in. More specifically, source freshness refers to the timeliness of the data in a source. A source is considered "fresh" if it has up-to-date data, while a source that is "not fresh" may have outdated data.

Source freshness in dbt is not intended to be a replacement for having good monitoring on your ingestion processes. You still need that. However, it is intended to complement that as an additional check. In some organizations, you may have separation of duties between the folks handling the data ingestion and the people using dbt. If the folks handling ingestion do not have good monitoring in place or not communicating with the people utilizing dbt, then it is really nice having an option like source freshness. We would still highly recommend finding ways to get monitoring in place and/or bridge the gap of communication.

Note Source freshness in dbt is not a replacement for having good monitoring on your ingestion processes.

Another benefit of source freshness is that if you run the check as the first step in your transformation load, then you can error out the rest from running. This is helpful when you want to avoid running your full transformation process when the data is stale. Even if you know there is an ingestion problem already, it is helpful to know that your job(s) will fail automatically early on if they continue to kick off based on a schedule. As soon as the ingestion issue is resolved, then your jobs will resume working without any manual intervention. We favor the approach of failing early and loudly when things aren't working as expected in our transformation process.

Configuring Source Freshness

To configure source freshness, you need to add some additional configuration to your sources.yml files. In Figure 3-2, we showed an example that included this and lightly touched on the topic, but let's dive into it some more. Let's start by just understanding the YAML syntax which is shown in Figure 3-3.

```
freshness:
warn_after:
  count: <positive_integer>
  period: minute | hour | day
error_after:
  count: <positive_integer>
  period: minute | hour | day
filter: <where-condition>
```

Figure 3-3. *Source freshness syntax*

Figure 3-4 shows another example of a sources.yml file with just source freshness configured.

```
1   version: 2
2
3   sources:
4     - name: Product
5       database: Product
6       freshness: # default settings
7         warn_after: {count: 24, period: hour}
8         error_after: {count: 48, period: hour}
9       loaded_at_field: DateLoaded
10
11      tables:
12        - name: items
13          freshness: # override default settings
14            warn_after: {count: 12, period: hour}
15            error_after: {count: 24, period: hour}
16
17        - name: purchases # use default freshness
18
19        - name: item_type
20          freshness: null # do not check for freshness
```

Figure 3-4. *Example sources.yml with source freshness configured*

In Figure 3-4, we have added the source freshness configuration in several ways to show some of the options you have. Let's start by just looking at the syntax for source freshness, and then we will look more at the options:

> **freshness**: This is the block used to let dbt know you are configuring source freshness.

warn_after: This tells dbt to trigger a warning message if the source has not been refreshed in the configured time. Warnings will not fail your build, but will trigger a message and write to the logs. To set up the time, you need to supply a period (minute, hour, day) and a count.

error_after: This tells dbt to trigger an error message if the source has not been refreshed in the configured time. Errors will fail your build until the issue is resolved. To set up the time, you need to supply a period (minute, hour, day) and a count.

loaded_at_field: This is the date/time column in your table that dbt uses to check for freshness. You can use any date/time column, but it is recommended that you use a metadata column specific to your ingestion process that tracks when the data was loaded.

filter: This is the place you can add a WHERE clause to your check to limit the amount of data scanned. This is particularly useful when you are trying to improve performance and this check is taking too long. Note, though, that this value only applies to the source freshness query and not anything else in the sources.yml file.

Now that you have an understanding of the configuration values, let's look back at Figure 3-4 at how we set this up. First, we set up a default configuration for the database (Product) that will be applied to all tables that we add. This default configuration says to use the DateLoaded field to check the freshness and warn if that date/time is older than 24 hours and error if that time is older than 48 hours. We don't have to add this at the database level and could have added it at the table level as well.

Note Source freshness can be configured at the database or table level.

The database level source freshness creates a default for everything else we add, but that doesn't mean that we can't override that. We can override the warning time, error time, check column, and even tell it to not check the freshness at all. To override it, we just need to specify the value at a more granular level in our YAML file. For example, if we had one table that we wanted stricter or looser settings on, we could specify values for that one table. In dbt, the more granular setting always takes precedence. So if you ever apply a default and want to override it for select objects, you can.

You have a lot of customization of how you handle this, and different teams/organizations will take different approaches. As authors of this book, we generally start with configuring source freshness at the table level and limiting it to essential tables that we know incur a high(er) volume of data changes. In most cases, the same process loading those tables is also what is loading others, so if there is an ingestion issue, we would know with just checking a few places. We also make sure we have good monitoring in place on our ingestion pipelines so we know there is a problem before dbt tells us.

Executing Source Freshness Checks

dbt does not automatically run your source freshness checks as part of **dbt build** or **dbt run**, instead they provide a separate command that handles this: **dbt source freshness**. You will want to make sure that this runs before you run your transformations if you want to make sure your data is up to date. You may also want to consider having a separate process that runs on a more frequent basis that is separate from your transformation process so you catch issues sooner.

If you are using dbt Cloud, then there is a checkbox that you can click when creating a new job that will automatically execute this (as seen in Figure 3-5), or you can just add a step in your process that executes **dbt source freshness**.

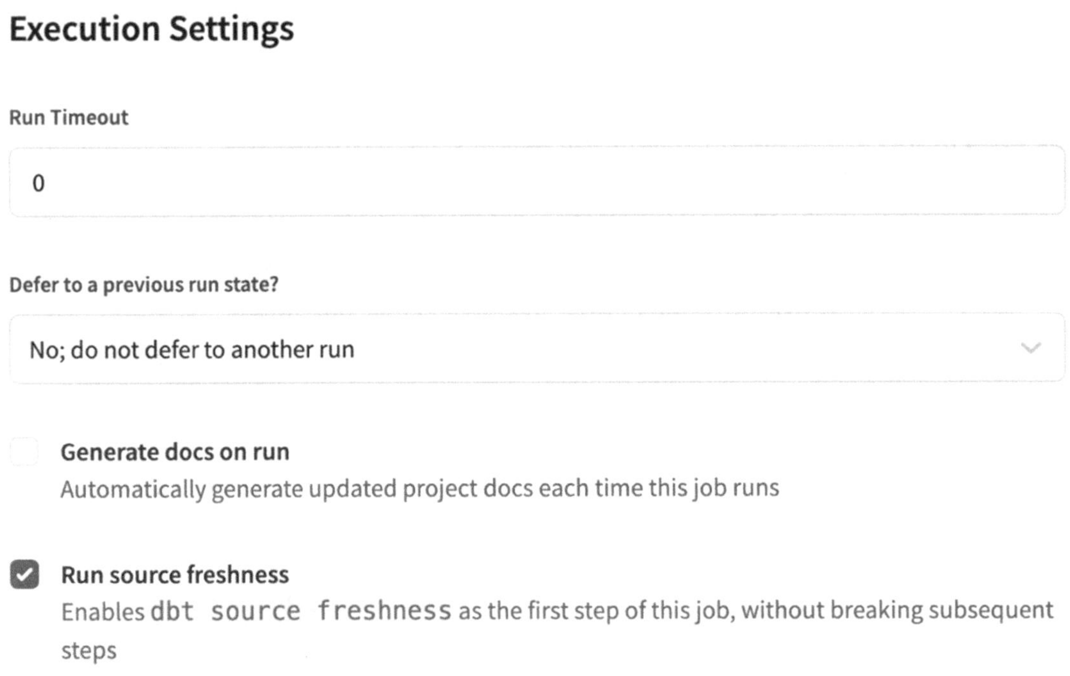

Execution Settings

Run Timeout

0

Defer to a previous run state?

No; do not defer to another run

Generate docs on run

Automatically generate updated project docs each time this job runs

Run source freshness

Enables `dbt source freshness` as the first step of this job, without breaking subsequent steps

Figure 3-5. *dbt Cloud checkbox to include source freshness*

Other Source File Options

We have covered the basics of what is needed to create your sources.yml file; however, if you look at the syntax presented earlier in the chapter, there are a few more options that we haven't covered. Some of these will be covered in more detail throughout the remaining chapters in the book:

> **loader**: Text description for the tool that loads the specified source data into your data store. This field is only used for documentation purposes.
>
> **meta**: The meta field is a way to attach text metadata to an object in dbt. This can be added to sources.yml files and can contain things like the data source owner, whether the source contains sensitive data, or the source system name. It can be used for anything you like, but is a way to attach metadata to your objects that extends beyond dbt's out-of-the-box coverage.

tags: These are optional metadata that you can attach to your models/objects that can then be used as part of the resource selection syntax. For example, when running a dbt command like *dbt run*, you can filter to running for or against models with tags.

quoting: This specifies whether dbt should quote databases, schemas, and identifiers when utilizing this source data. dbt automatically sets this to default values based on your data store. It is generally recommended to not change this unless you have a specific reason to do so.

Utilizing Sources in Multiple Environments

If you plan to create multiple environments, and you absolutely should, for your warehouse (i.e., Dev, QA, Prod, etc.) and need to be able to support that with your dbt project, we highly recommend considering using environment variables in your sources.yml file for anything that could change from environment to environment. That means if your database or schema names are different between your development and any deployment environments, then you will want to consider this; otherwise, your project will break when you promote your code. To reference an environment variable, you will utilize the env_var function in your YAML files. Environment variables can be used in your profiles.yml, dbt_project.yml, and sources.yml files.

We often see Snowflake users utilize the same instance for all of their environments since there is separation of storage and compute. Teams are able to utilize separate warehouses (compute buckets) between environments and specific roles to limit access without the additional complexity of running separate instances. When doing so though, you can only utilize a single database name once which creates naming variance between environments. This is where environment variables come into play. By adding a variable into your sources.yml file, dbt can interpret the values from your project based on which environment you are running in.

We will cover this topic in a lot greater detail in Chapter 10 of this book, but we wanted to go ahead and mention it here so you are aware of what is possible.

Generating Your Source File

When generating your sources.yml, you have a couple of options. You can type everything in manually, or you can programmatically create it. If you have a lot of sources and objects you need to reference, this can take a while. We believe the time to manually populate can be worthwhile because it provides an opportunity to add descriptions, source freshness, tests, etc. to your sources.yml file that you may have otherwise put off. Just remember that you only have to populate this once.

If you have a lot of sources with lots of tables, then generating your sources.yml can be a lengthy process. Fortunately for us, we can utilize the codegen package found on the dbt package hub (*hub.getdbt.com*) to generate most of the scaffolding for us. There is a macro in this package called generate_source that generates lightweight YAML for a source which you can copy and paste into your sources.yml file. Because this is an open source package that may change between now and the time you are reading this, we are not going to provide specific steps on how to use it. However, it is something that we believe will always exist, and we recommend checking out the latest package version on the dbt package hub for instructions. The codegen package is a staple in the dbt arsenal and is one that we also cover throughout the book, primarily in Chapter 6.

What Are Seeds?

Seeds are comma-separated values (CSV) files in your dbt project that dbt can load into your data store. They are best suited for small unchanging datasets that you may not have a source system to pull from. Seeds are stored, version controlled, and code reviewable in your dbt project, whereas other source data is maintained outside of it. Good use cases for these would be mapping tables like country codes to country names, US state abbreviations to US state names, or some other type of static mapping or creating lists of things you need to manually keep track of to include or exclude from lists. Bad use cases would be large changing datasets, loading raw data that was exported to CSVs, or any kind of data containing sensitive information.

Here are a few things to consider when deciding whether to use seeds in your dbt project:

Data freshness: Seeds are typically used to insert a set of data into a database and are not meant to be regularly modified or updated. This means that the data in a seed is not likely to change much over time. If you need to regularly update the data, then seeds may not be the best option, as you would create a lot of additional manual work.

Data volume: Seeds are intended for small amounts of initial data that is needed to set up a database. If you have a large amount of data that needs to be loaded into your database on a regular basis, seeds may not be the most efficient option. Instead, you may want to consider using another option. As a general rule of thumb, we believe seeds should typically be less than 5000 rows. Additionally, when seed files exceed 1MB, then dbt is unable to do checksum state comparison.

Data dependencies: Seeds are often used to set up reference data that is needed by downstream models. If your models depend on data that changes frequently or is updated on a regular basis, seeds are not the best option.

In dbt, seeds are typically stored in the seeds directory of your dbt project and get materialized as tables in your warehouse. There is a separate command that you need to run to populate your tables with seed data called **dbt seed**. When you run dbt seed, dbt will read from the seed directory and create your tables. dbt build will automatically build your seeds as well when it runs.

Let's take a look at an example seed file called USStates.csv that maps state abbreviations to state names. Figure 3-6 shows a partial look of what this file looks like.

```
1   STATEID,STATENAME,STATEABBREV
2   1,Alabama,AL
3   2,Alaska,AK
4   3,Arizona,AZ
5   4,Arkansas,AR
6   5,California,CA
7   6,Colorado,CO
8   7,Connecticut,CT
9   8,Delaware,DE
10  9,District of Columbia,DC
11  10,Florida,FL
12  11,Georgia,GA
13  12,Hawaii,HI
14  13,Idaho,ID
15  14,Illinois,IL
16  .... continued
```

Figure 3-6. *Example of USStates.csv seed file*

With a seed file, the first line will be the headers, and each subsequent line will contain your data. The file is structured the same as you would see in any text version of a CSV file. It is important that files are formatted correctly, or you could get an error when trying to build. The header line is where we commonly see issues when there are problems and is usually related to formatting or a mismatch of the headers with the number of columns.

When the seed is generated, it will materialize in your warehouse as a table named the same as the file name. There are configuration options to change this behavior if you want, but not directly in the file itself like other models. We recommend just naming the CSV file what you would like the table to be called in your warehouse. It's worth noting that dbt will rebuild the tables associated with your seed files each time you invoke **dbt seed** or **dbt build**. Seeds function very similarly to models in your project and are referenced using the ref() Jinja function. We will cover models and the ref() function in much greater detail in the next chapter.

Executing Seeds

To build your seed data in your data warehouse, you just need to run **dbt seed**. This command will load all CSV files located in the seeds-path directory of your dbt project. Like the dbt run command, you also can add the --select flag to limit it to only building a specific seed file.

Using Figure 3-6, we could execute one of the following commands to build our list of US states in our warehouse. The first option shows us running the base command, and the second option shows us running just the specific seed:

```
-- Run all seeds including USStates.csv
dbt seed
-- Run just the USStates.csv seed
dbt seed --select USStates.csv
```

The dbt build command also automatically builds seeds for you, so you do not have to run the dbt seed command in conjunction with that. As always, dbt builds your DAG for you and will build the seeds before any downstream model is able to use them.

Summary

In this chapter, we looked at the important role sources play in your dbt project. We learned how advantageous it is to manage these outside of your transformation models and how easy it is to update when a source change occurs. Sources are contained in their own YAML file and are very configurable with lots of additional options such as descriptions, tests, and source freshness. Your sources.yml file does have to be manually created and populated; however, dbt Labs has an open source package called codegen that can help here.

We also looked at the role that seeds play in our dbt project. Seeds are CSV files that we can create that are great for small unchanging data that may not be part of our source data. Seed data is stored in our dbt project and is source and version controlled. Seeds are treated very similarly to the way we treat models. In the next chapter, we will dive into models which foundationally are one of the most important aspects of utilizing dbt.

CHAPTER 4

Models

In the not so distant past, most data teams fully relied on data transformation tools that were built around drag-and-drop graphical user interfaces (GUIs) or stored procedures. Unfortunately, implementations of these solutions tend to end up being very complex and in a consistently unreliable state. One of the main reasons for this is that software engineering best practices, which are also applicable to data teams, are not used. For example, data transformations tend to not be modular, data quality testing is infrequently implemented, and version control is not used.

In addition to these issues, teams using stored procedures or a built in-house transformation tool will be responsible for writing boilerplate code to manage the objects in the data warehouse. For example, you would need to write the DDL to create a table and the DML to update that table in future runs of your transformation pipeline. Fortunately, dbt abstracts away the boilerplate code and empowers you to make up for the downfalls of other transformation tools and processes as it relates to software engineering best practices.

In this chapter, we will discuss the different types of models that you can implement with dbt; as of the writing of this book, dbt offers up the ability to write models using SQL and Python. We will then look at the types of materializations that are available for both SQL and Python models, model configurations, and the commands that can be used to run models. In the first several sections of this chapter, we will be walking you through examples of building models, and we will do so using the staging-intermediate-marts pattern that is commonly used in dbt projects. However, as we've mentioned time and time again, dbt is simply a tool to help you with your data transformations, and how you structure those transformations is entirely up to you. In these examples, you should mainly pay attention to the syntax and worry less about the structure of the project. Toward the end of the chapter, we will discuss modularity within your transformations, and within this section we will discuss a few ways that you might structure your project.

C. Cyr and D. Dorsey, *Unlocking dbt*, https://doi.org/10.1007/978-1-4842-9703-2_4

SQL Models

Building models is, in our opinion, the single most important aspect of dbt that you should understand. Having a sound understanding of SQL models in dbt sets the stage for the remainder of this book because each of the remaining chapters of this book will build upon this knowledge. For example, in Chapter 6 we will discuss Jinja and macros, which are frequently used to add Python-like control flows to the compilation step of your SQL models. Or in Chapter 8, we will discuss tests and how to implement them to check the quality of the output data of your models.

Today, dbt offers its users the ability to write models in both SQL and Python, but by and large SQL still dominates both dbt and the entire data industry as the de facto language for data transformations. Later in the chapter, we will discuss why dbt added support for Python and when it makes sense to build your models using Python, but for now let's focus on SQL. As of the writing of this book, we are firm believers that, for most people, writing models in SQL is the place you should begin your journey of learning about dbt models. We think this way for a few reasons, including

- All dbt adapters support SQL models.
- They tend to be the better option for simple transformations.
- Extremely low barrier to entry because everything is a select statement.
- SQL is historically more widely adopted than Python by data professionals.

In short, we believe that for most people SQL should be your default language of choice when building dbt models, and you should use Python to build models when the transformation is complex and could more eloquently be handled with Python syntax. We will discuss what some of those use cases are later in the chapter, but for now let's discuss the materialization types that are available for SQL models in dbt.

Recall from Chapter 1 that we introduced you to the concept of **materializations** in dbt, which can be defined as the method that dbt will use to generate models and write them, or not write them, to your database. There are four materializations that can be used with SQL models:

- View
- Table
- Incremental
- Ephemeral

For the remainder of this section, we will discuss considerations you should take when choosing a materialization and walk you through examples of how to implement each. If you want to follow along with these examples, be sure that you have completed the project setup process discussed in Chapter 2, and be sure that you have properly seeded your database with the .csv files that have been included in the book's GitHub repository: `https://github.com/apress/unlocking-dbt`. If you still need to seed your database, as long as you have set up your dbt project correctly and checked out the book's git repo, then all that you need to do is execute the `dbt seed` command. By running this command, you will end up with four tables in your database that we will use to build transformations for the rest of the book:

- raw_products
- raw_orders
- raw_orderitems
- raw_customers

View (Plus, Model Basics)

The first materialization that we will discuss is the view materialization. When starting out with dbt, the view materialization is the easiest to understand for most people that have a background in SQL because writing a view in dbt is very similar to writing a view in raw SQL. In fact, this materialization generates a view in your database for you based on the SQL select statement that you write. Later in this section, we will walk you through an example of building a view in dbt and how it differs from writing a view in raw SQL. This example will also include a more broad explanation of how dbt models differ from writing vanilla SQL. But, first let's examine when you might want to use the view materialization.

Caution The view materialization should not be confused with materialized views. The view materialization in dbt simply creates a standard view object in the database, whereas materialized views are database objects that persist data. We won't discuss materialized views in the book because dbt doesn't currently support this type of database object.

As you may know, a view is a query that is stored in a database as a queryable object. A view can also be thought of as a virtual table because no data is persisted within the view object. As such, every time a person queries a view, the database must run the entire underlying query. As you can imagine, if you're working with a complex transformation this has the potential to introduce performance issues. In Table 4-1, we list advantages and disadvantages of using the view materialization.

Table 4-1. *View materialization advantages and disadvantages*

Advantages	Disadvantages
Simple to create	Performance can decline quickly
Minimal storage is occupied in the database	Can be difficult to maintain if the source table(s) has unpredictable schema changes
The latest data is always available	Not ideal for complex transformations

Considering the advantages and disadvantages from Table 4-1, we believe that you should use the view materialization mindfully, and not simply as the default. This goes against the grain of dbt's official recommendation in their documentation (`https://docs.getdbt.com/docs/build/materializations`), which is to typically build your models using this materialization at first and move to a more complex materialization if performance improvements are needed. This is sound advice when you are first starting out with building dbt models, but as you become more experienced with building models and are comfortable with the optimization patterns of your database, this recommendation becomes an oversimplification. We're of the opinion that when you start to build a new model, if you know that a view will not provide sufficient performance then you shouldn't start with the view materialization, but instead should look at one of dbt's other materialization options, such as the table materialization.

Although we don't agree that the view materialization should always be your default, if you don't believe performance will decline by using the view materialization then it may be a good place to start. You don't want to fall into the trap of overoptimization at an early stage of development. Keep it simple, and you will thank yourself later (maybe your coworkers too).

For now, let's walk you through how to build a dbt model using the view materialization. For this example, we will be building a model on top of the *orders* data that was used to seed the database. Let's first look at how we would build this view in the traditional way using raw SQL and DDL (data definition language) for the boilerplate code. This will provide the baseline for what we are trying to achieve: building a view in the database. Once this is done, we can then see what changes would need to be made to convert this to a dbt model.

In Listing 4-1, we've provided the SQL that would traditionally be needed to be written to create a view. In this example, the view is being created by using the statement `create or replace view as` so that the view can be generated as the result of a `select` statement. This select statement is quite simple as it doesn't contain any joins and is conducting a few very lightweight transformations on some of the columns such as casting timestamps to different formats and reformatting people's names.

Listing 4-1. Example of creating a view using raw SQL

```
create or replace view dbt_learning.public.stg_furniture_mart_orders as
select
  ord.OrderId,
  ord.CustomerId,
  initcap(ord.SalesPerson) as SalesPerson,
  cast(ord.OrderPlacedTimestamp as timestamp_ntz) as OrderPlacedTimestamp,
  ord.OrderStatus,
  cast(ord.UpdatedAt as timestamp_ntz) as UpdatedAt
from dbt_learning.public.raw_orders as ord
```

With just a few minor changes to the code in Listing 4-1, we can easily convert this to be a dbt model that will be materialized as a view; this can be seen in Listing 4-2.

Listing 4-2. Example of configuring a view model

```
{{
  config(
    materialized='view'
  )
}}

select
  ord.OrderId,
  ord.CustomerId,
  initcap(ord.SalesPerson) as SalesPerson,
  cast(ord.OrderPlacedTimestamp as timestamp_ntz) as OrderPlacedTimestamp,
  ord.OrderStatus,
  cast(ord.UpdatedAt as timestamp_ntz) as UpdatedAt
from {{ ref('raw_orders') }} as ord
```

If you compare Listing 4-1 to Listing 4-2, there are likely three things that stand out to you immediately about the dbt model:

1. There is some configuration code at the top of the file.
2. There is no `create or replace view as` statement.
3. Instead of selecting from the table directly, the dbt model is selecting from `{{ ref('raw_orders') }}`.

Starting at the top of the file, you likely noticed a small bit of configuration code:

```
{{
  config(
    materialized='view'
  )
}}
```

This is known as the *config block* and is used within .sql files in a dbt project to define configurations for an individual model. As you can see, we have set one configuration, `materialized`, to equal `'view'`. As you likely guessed, this is how we instruct dbt to use the select statement within this file to generate a view in our database. In fact, this provides the rationalization for why we don't need to include a `create or replace` `view as` statement in our SQL file. Since we have configured this model to be a view, dbt will inject the boilerplate DDL for us at runtime. As such, to get started building models, you only need to write select statements for any of your SQL models, set the materialized configuration, and dbt will handle the rest for you.

Tip The view materialization is the default in dbt, and as such you don't actually need to set the configuration, like done earlier. But, we have done it here just to show you how the configuration is used, and in our opinion, it never hurts to be explicit when writing code.

The final difference that you likely noticed between Listings 4-1 and 4-2 was that in the original example the `raw_orders` database object was referenced directly:

```
select
...
from dbt_learning.public.raw_orders as ord
```

However, in the latter example (Listing 4-2), we instead use the `ref` Jinja function to reference the database object that we are selecting from in this model. In Chapter 3, we introduced you to the `source` Jinja function, which is used to directly reference untransformed objects in your database, otherwise known as *sources*. The `ref` function works similarly, but instead of referencing a source database object, it is used to reference an upstream dbt transformation, including

- Other models
- Snapshots
- Seeds

If you recall from the beginning of the chapter, we asked you to seed your database using the dbt seed command, so that the .csv files that were included in the book's GitHub repository were materialized as tables in your target database. Since we created those seeds, we are able to use them in downstream models, such as the one in Listing 4-2, using the `ref` function. Of course, even if you aren't using the provided code and seed data, this would still hold true to your project as long as you have seed data or sources to reference. Next, take note of the string that we are using within the `ref` function:

```
from {{ ref('raw_orders') }} as ord
```

Since we are referencing a seed file named raw_orders.csv, dbt allows us to reference them directly based on the name of the file. For now, we are only working with seeds directly using the ref function, but the same logic holds true when you start building models based off of the output of other upstream models. For example, if you have a model file named my_first_dbt_model.sql and you want to reference it in a downstream model with a file name my_second_dbt_model, you would do so again by using the `ref` function: `{{ ref('my_first_dbt_model') }}`. You will want to remember that by default dbt allows you to reference upstream models, seeds, and snapshots in this way because you will do it a lot when you develop with dbt! Hopefully, this provides you an understanding of how to use the ref function, but why bother using it in the first place? If you recall from Chapter 3, we discussed how by using the source function dbt will be able to infer relationships between your dbt assets. Well, the same holds true when using the ref function, and dbt is able to work its magic to know the order in which to build your models, seeds, and snapshots. Additionally, using the ref function provides the context needed by dbt to generate a DAG that can be visually represented in your documentation. If you were to hardcode database object names instead of using the ref and source Jinja functions, then dbt wouldn't be able to be aware of the relationships between your assets. For that reason, we recommend you *never* hardcode database object names directly in your transformation code.

While not directly related to the view materialization, we do want to close out this section with a few more model basics that will help you further understand how dbt takes models that you have defined in .sql files, compiles them, and runs them. Let's first start with understanding how dbt will convert the code within model files that contain configurations and Jinja functions into raw SQL that the database will understand. The first step that dbt takes is to *compile* the code, and in this step the config block will be removed from the compiled SQL and ref (or source) Jinja functions will be removed and replaced with database objects. Ephemeral models are an exception to this that we will discuss shortly, but aren't something you should worry about at this point.

To dissect what dbt is doing under the hood, you can run the command `dbt run --select stg_furniture_mart_orders`, and then we can explore the compiled and run files that get generated. But first, let's quickly cover what this command is doing; this command is telling dbt that you want to run the model within the file: ~/models/staging/furniture_mart/stg_furniture_mart_orders.sql. The `dbt run` command is used to run models, and as such dbt searches the models directory, and its subdirectories, for a file matching the name stg_furniture_mart_orders. Keep in mind that file names need to be globally unique because dbt would be unable to automatically infer model names otherwise. As such, if your file names are not unique, you will run into errors.

After running the preceding command, you can navigate to the *target* directory in your dbt project. The target directory is used by dbt to store run artifacts. In Chapter 1, we discussed what this directory is used for, but as a quick reminder, it is generated automatically by dbt whenever you run a command. The target directory is used to store a multitude of metadata, compiled SQL, and run SQL (the code actually shipped to your database). For more details on what each of the artifacts is in this directory, we'll refer you back to Chapter 1. Throughout the book, we will look at the different files in this directory, but for now we will be focusing on the files within the *compiled* and *run* subdirectories. If you take a look within the *compiled* subdirectory, you should see a file named *stg_furniture_mart_orders.sql*. As you may have guessed, this file houses the compiled SQL resulting from our model file. This corresponds directly to the name of our file in the models directory where we defined this transformation. Of course, you could have named this transformation anything; this is simply what we named it in the example we provided you. Take a look at Listing 4-3 to see how dbt will convert our model file to plain SQL.

Listing 4-3. Example of **`compiled`** dbt SQL code

```
select
  ord.OrderId,
  ord.CustomerId,
  initcap(ord.SalesPerson) as SalesPerson,
  cast(ord.OrderPlacedTimestamp as timestamp_ntz) as OrderPlacedTimestamp,
  ord.OrderStatus,
  cast(ord.UpdatedAt as timestamp_ntz) as UpdatedAt
from dbt_learning.public.raw_orders as ord
```

This is a very simple example, but as you can see, the config block has been removed from the top of the file, and the ref function has actually been converted to a fully qualified database object name. While this is interesting, it doesn't paint the full picture because this is still only a select statement, but we instructed dbt to generate a view in the database. This may make you wonder, what did dbt do with the config block and how will this select statement create a view? We would expect that there is also some DDL, such as a create statement, preceding the select statement. Well, if we shift our focus away from the *compiled* subdirectory and look within the *run* subdirectory, we should again find a file named *stg_furniture_mart_orders.sql* (or whatever you chose to name your model).

If you were to open up this file, also shown in Listing 4-4, you will see how everything comes together. Notice that the compiled SQL query has been injected into some DDL to create, or replace, the view. As we discussed earlier, this happens because we set the materialization configuration to `'view'`.

Listing 4-4. Example of **`run`** dbt SQL code

```
create or replace view dbt_learning.public.stg_furniture_mart_orders
  as (
select
  ord.OrderId,
  ord.CustomerId,
  initcap(ord.SalesPerson) as SalesPerson,
  cast(ord.OrderPlacedTimestamp as timestamp_ntz) as OrderPlacedTimestamp,
  ord.OrderStatus,
  cast(ord.UpdatedAt as timestamp_ntz) as UpdatedAt
from dbt_learning.public.raw_orders as ord
);
```

While this example is very simple, as you can imagine dbt can save you time and effort when building your data transformations because it manages most of the boilerplate code for you. We will see more complex examples as the chapter progresses, but we didn't want to throw you straight into the deep end when learning the basics. We can summarize what happens under the hood when you run a dbt model, as we saw earlier, as

1. Model files are *compiled* and converted to plain SQL.
2. The compiled SQL is injected into the appropriate boilerplate code (in this case, dbt added code to create a view).
3. The compiled SQL with boilerplate code included (stored in the *run* folder) is shipped off to build an object in your data warehouse.

This process doesn't only happen for the view materialization, but all other materialization types as well. Let's now take a look at how the table materialization works and when you may choose to use it instead of the view materialization.

Table

There have been many times when we were using a view as part of our data transformation process, or maybe using a view to expose a query to end users, and performance degraded over time because the data behind the view grew larger or the logic in the view grew exceedingly complex. Maybe you have been in a similar scenario as well. In this section, we will discuss the table materialization, which is the next logical progression in dbt materializations when the view materialization starts to show underwhelming performance.

As the name implies, this materialization will build a table in your database based on the select statement that you have written in your model file and will either create or replace the table each time that you invoke dbt against the model. Being that this materialization creates a table in your database, you will likely experience improved performance, in comparison to a view, when retrieving data from it. That is because tables actually store the data, and not a reference to it like views do.

Note We will regularly make reference to "invoking dbt" or "executing dbt." What we mean by this is the act of submitting a command to run one or more models, seeds, or snapshots. Some examples of this would include the commands: `dbt run`, `dbt build`, `dbt snapshot`, and `dbt seed`.

While we aren't going to get too far into the particulars of how database tables work, we do think it is worth noting the difference between *column stores* and *row stores* and the implications of each for analytics workloads. As the name suggests, a row store table stores blocks of data based on rows. Row store tables are frequent in Online Transaction Processing (OLTP) databases because of the ability for these tables to quickly retrieve and operate on a single row of data. This is useful for transaction-based systems because they are typically working on a single or small subset rows, and the transaction needs access to the entire row of data that it is operating on so that it can appropriately lock the record and make updates without conflict. However, row-based tables can have underwhelming performance for analytics workloads because of the need to return all of the data in each row that is selected.

Fortunately, many modern data platforms that are built for analytics, or Online Analytical Processing (OLAP) databases, don't use row store tables, but instead they use column store tables. In a column store table, data is stored based on columns instead of rows. This enables what is known as column compression. With column compression, data is stored in individual columns, and so each storage block shares the same data type, and storage can be optimized for individual column retrieval. A columnar store is beneficial to analytics systems because data lookups happen much quicker, but this comes with the trade-off of slower data insert and updates. However, having faster column retrieval is advantageous in the world of analytics because it is common to only select a few columns and run an aggregation against them. If you are interested in a more in-depth coverage of row stores and column stores, we recommend that you check out the book *Fundamentals of Data Engineering* by Reis J and Housley M (2022).

Now, let's cover how to build a dbt model using the table materialization. This example can be found in the book's GitHub repository at *~/models/intermediate/operations/int_orders_items_products_joined.sql*. As always, you can follow along using the provided repo if you want to, but if you'd rather not, that's fine too. The concepts that we will discuss are transferable and can be easily applied to dbt models you build on your own.

In Listing 4-5, we provide an example of a model where we take an intermediate step to join two of our staging models together. While in our project neither of these objects have very much data in them, imagine a real-world scenario where these tables contain billions of records. If this were true, it could be useful to materialize these two joined models as a table, so that downstream we can just join to this one table instead of a view that requires this join to rerun each time.

Listing 4-5. Example of materializing a model as a table

```
{{
  config(
    materialized = 'table'
  )
}}

select
  itm.OrderItemsId,
  itm.OrderId,
  pro.Product,
```

```
  pro.Department,
  pro.Price
from {{ ref('stg_furniture_mart_orderitems') }} as itm
join {{ ref('stg_furniture_mart_products') }} as pro
  on itm.ProductId = pro.ProductId
```

As you can see in Listing 4-5, the materialization configuration is still set using the config block at the top of the model file, which is exactly the same as how we configured a model to be a view except this time we set the configuration to equal `'table'`. Other than that, all that needs to be done is to write a select statement to return that data that should be materialized in the destination table. If you recall from the prior section, when you execute a model two main things happen: the query is compiled and then boilerplate code is added, so the object can be built in your database. The concept of compiling is pretty straightforward, so let's skip ahead and look at the SQL that dbt will generate to be shipped to your database.

Notice that in the first line of Listing 4-6, dbt injected some DDL to tell the database to create or replace a table using a select query. More specifically, dbt wraps the compiled select statement in a `create or replace table` statement.

Listing 4-6. Example of SQL that will be executed against your database by dbt when using the table materialization

```
create or replace table dbt_learning.public.int_orders_items_products_
joined as
(
select
  itm.OrderItemsId,
  itm.OrderId,
  pro.Product,
  pro.Department,
  pro.Price
from public_staging.stg_furniture_mart_orderitems as itm
join public_staging.stg_furniture_mart_products as pro
  on itm.ProductId = pro.ProductId
);
```

Also, note that dbt is creating a fully qualified database object of `dbt_learning.public.int_orders_items_products_joined`. Of course, the name of the table corresponds directly to the name of the .sql file where the model is stored, but also notice that it has chosen to place the table in the `dbt_learning` database and the `public` schema. This is because of how we defined the **target database** when we set up our dbt project. We told dbt to always materialize objects in this database and schema. Later in the chapter, we will talk more about model configuration and how you can change this behavior some.

We mentioned this earlier in the section, but it is worth mentioning again that the table materialization will rebuild the entire table in your database each time you invoke dbt against the corresponding model. While this is going to be more performant than a view for end users, if you are working with large volumes of data, rebuilding tables every day may not be realistic because of how long it would take to process the data. In the next section, we will take a look at a materialization that serves to solve this problem by incrementally loading data into your destination tables.

Incremental

So far, the two materialization types that we've covered, table and view, take the approach of recreating objects in your database each time the dbt is run. While this can work well for teams transforming low volumes of data, or teams that have an infinite cloud budget, there are times when data needs to be incrementally processed so that pipelines can deliver data in a reasonable amount of time. *A reasonable amount of time* is arbitrary and up for you and your team to decide, but if you aren't able to process your data within that time frame, whatever it may be, you may want to consider the incremental materialization.

What do we mean by *incrementally processed* anyway? When we talk about incrementally processing data, we are referring to only transforming the data that hasn't already been transformed and then adding this newly transformed data to our target table. For example, suppose that you have a source table that stores data related to orders from an ecommerce website, and we need to transform that data and store it in a table for the analytics team to operate on. But, this table has billions of records, so recreating the target table each time dbt is invoked is unrealistic because it takes hours to build. Instead, we can build the table once and then, in future invocations of dbt, only transform any data that is new since the prior run and then **upsert** this data to the target table. For the rest of this section, we will talk about the different ways to implement incremental transformations in dbt and when to use different patterns and strategies.

As with all other materialization types, incremental models are defined by writing a select statement and setting configurations to tell dbt that the model should be materialized incrementally. With that said, using the incremental materialization is more complex than the view or table materialization type, and as such you should only move to the incremental approach when the optimization benefits outweigh the cost of development and maintenance. We typically move to the incremental materialization when the normal table materialization can no longer keep up with our pipeline SLAs.

Understanding the syntax of incremental models can take some getting used to, especially if you come from the old world of using stored procedures to load tables incrementally. Before dbt, if you wanted to load a table incrementally following a data transformation, you might have followed a pattern similar to

1. Determine the state of the destination table. That is, when was the last time data was loaded into it?
2. Transform new data and insert it into a temporary table.
3. Merge, or upsert, the data from the temporary table into the destination table.

Fortunately, using incremental models simplifies this process, and we only have to worry about writing the select statement to transform the data. No need to store it in a temporary table, and no need to write the merge/upsert logic. As with most things, dbt will handle all of the boilerplate code for you.

Caution The following examples are configured to run against Snowflake, and by default the dbt-snowflake adapter uses a **merge** strategy to build incremental models. As such, in the next few examples, we will discuss incremental models from this perspective. If you are using a database other than Snowflake, please check the documentation of your database's dbt adapter to understand which incremental strategy it uses. This will provide you a baseline understanding of incremental models, but later in the chapter, we will discuss each of the incremental strategies that dbt offers and how they are configured.

Let's start with a simple example of how you can implement a model in dbt using the incremental materialization. This example will generate a model that will store data related to the revenue of an order. If you want to follow along, the example can be found at this path of the book's repo, *~/models/marts/finance/fct_revenue_orders.sql,* or can be seen in Listing 4-7.

Listing 4-7. Example of a basic incremental dbt model

```
{{
  config(
    materialized = 'incremental',
    unique_key = 'OrderId'
  )
}}

with get_orders_revenue as(
  select
    pro.OrderId,
    sum(pro.Price) as Revenue
  from {{ ref('int_orders_items_products_joined') }} as pro
  group by 1
)

select
  ord.OrderId,
  ord.OrderPlacedTimestamp,
  ord.UpdatedAt,
  ord.OrderStatus,
  ord.SalesPerson,
  rev.Revenue
from get_orders_revenue as rev
join {{ ref('stg_furniture_mart_orders') }} as ord
  on rev.OrderId = ord.OrderId
```

This example starts off with a config block at the top of the model similar to other model examples that we have covered within this chapter. Within the config block, the materialization configuration is set to `'incremental'`, but notice that there is an additional configuration, `unique_key`. This configuration is used by dbt to determine which column makes up the grain of the table and will use that column to conduct the merge, or upsert, operation to insert new records and update existing ones in the destination table. In Listing 4-7, the `unique_key` is set to `OrderId`, and as such dbt will expect that our select query returns only one row per `OrderId`.

We do want to call out that dbt doesn't require the `unique_key` configuration to be set for incremental models. If the `unique_key` configuration isn't set, then dbt will insert every record returned from your select statement instead of running a merge/upsert operation. Most of the time, this isn't a desirable behavior because insert statements are inherently not idempotent, so we recommend always including the unique_key configuration unless you have a solid reason to not include it.

Definition Idempotence, as it relates to data pipelines, is the idea that inputs should always yield the same outputs regardless of when or how many times a pipeline has been run. As such, insert operations (without a corresponding delete) tend to not be idempotent.

With the exception of the config block, so far this model is set up just like all of the other model examples that we've covered in this chapter, but let's take a look at what actually happens when you run dbt using the incremental materialization.

On the initial run, dbt will run a statement to create your table, which can be seen in Listing 4-8. Of course, this is necessary because if there isn't an existing table, then dbt will have nothing to build into incrementally.

Listing 4-8. Query executed against your database by dbt on the first run of an incremental model

```
create or replace transient table dbt_learning.public.fct_revenue_orders as
(
with get_orders_revenue as(
  select
    pro.OrderId,
    sum(pro.Price) as Revenue
  from public.int_orders_items_products_joined as pro
  group by 1
)
```

```
select
  ord.OrderId,
  ord.OrderPlacedTimestamp,
  ord.UpdatedAt,
  ord.OrderStatus,
  ord.SalesPerson,
  rev.Revenue
from get_orders_revenue as rev
join public.stg_furniture_mart_orders as ord
  on rev.OrderId = ord.OrderId
);
```

However, on subsequent runs, dbt will trigger a different process so that records are inserted incrementally instead of running a full rebuild of the target table. But, before we dissect what commands are run against your database during an incremental build, you might be curious how dbt is aware that it is in fact an incremental build.

On every invocation of dbt, not just for incremental models, a set of commands will be sent to your database to understand what objects already exist. These commands vary depending on what target database you have configured dbt to use, but for Snowflake (which we use in our examples), dbt will run a compilation of the following commands:

- show terse objects in database <target.database>
- show terse objects in <target.database>.<target.schema>
- describe table <target.database>.<target.schema>.<some_table_name>

While these commands are specific to the Snowflake adapter, for all adapters dbt is effectively grabbing the metadata in your database to understand what objects already exist. By doing this, dbt is able to cache the metadata for the duration of the run so that it can determine actions to automatically do for you such as

- Generating schemas
- Building incremental models for the first time or incrementally
- Building snapshots for the first time or incrementally

Knowing that dbt grabs this metadata at the start of each invocation, on a subsequent run dbt will be aware that the `fct_revenue_orders` table already exists and as such is able to infer that this must be an incremental build of the model. So, instead of running a create or replace statement for this table, dbt will do the following:

- Create a temporary view using the SQL defined in the model file
- Run a merge statement to merge the results of the view into the target table
- Drop the temporary view

You can see the commands that dbt will execute for these three steps in Listing 4-9.

Listing 4-9. Example of the commands that dbt executes on subsequent runs of an incremental model

```
##Step 1
create or replace  temporary view dbt_learning.public_finance.fct_revenue_
orders__dbt_tmp as (
with get_orders_revenue as(
  select
    pro.OrderId,
    sum(pro.Price) as Revenue
  from public_staging.int_orders_items_products_joined as pro
  group by 1
)

select
  ord.OrderId,
  ord.OrderPlacedTimestamp,
  ord.UpdatedAt,
  ord.OrderStatus,
  ord.SalesPerson,
  rev.Revenue
from get_orders_revenue as rev
join public_staging.stg_furniture_mart_orders as ord
  on rev.OrderId = ord.OrderId
);
```

```
##Step 2
merge into dbt_learning.public_finance.fct_revenue_orders as
DBT_INTERNAL_DEST
using dbt_learning.public_finance.fct_revenue_orders__dbt_tmp as
DBT_INTERNAL_SOURCE
  on (DBT_INTERNAL_SOURCE.OrderId = DBT_INTERNAL_DEST.OrderId)
    when matched then update set
        "ORDERID" = DBT_INTERNAL_SOURCE."ORDERID",
"ORDERPLACEDTIMESTAMP" = DBT_INTERNAL_SOURCE."ORDERPLACEDTIMESTAMP",
"UPDATEDAT" = DBT_INTERNAL_SOURCE."UPDATEDAT",
"ORDERSTATUS" = DBT_INTERNAL_SOURCE."ORDERSTATUS",
"SALESPERSON" = DBT_INTERNAL_SOURCE."SALESPERSON",
"REVENUE" = DBT_INTERNAL_SOURCE."REVENUE"
    when not matched then insert
        ("ORDERID", "ORDERPLACEDTIMESTAMP", "UPDATEDAT", "ORDERSTATUS",
        "SALESPERSON", "REVENUE")
    values
        ("ORDERID", "ORDERPLACEDTIMESTAMP", "UPDATEDAT", "ORDERSTATUS",
        "SALESPERSON", "REVENUE")
;

##Step 3
drop view if exists dbt_learning.public_finance.fct_revenue_orders__dbt_
tmp cascade
```

The steps shown in Listing 4-9 are specific to the Snowflake adapter, so they may vary slightly for other adapters. However, the general concept will stay the same in the sense that dbt will handle the merge/upsert logic for you. Regardless, the need to understand how dbt builds models incrementally has been abstracted away, so as a user of dbt you don't really need to worry about this. We are only covering this information because we figured if you're reading this book, you were interested in a deeper dive of what happens behind the scenes than what the standard dbt documentation provides you.

Before we move on, we want to bring to your attention that in this example we have done nothing to limit the data that is transformed on each dbt run. As such, each time this model executes, all of the underlying data that is being queried by the model will be transformed, and then a merge operation will upsert that data to the target table.

It would be significantly more performant to understand the state of the target table prior to selecting new data to upsert into the target table. As with most things, dbt has provided the necessary features to solve the problem of limiting data retrieval on incremental builds.

Limiting Data Retrieval

Often when running data transformations that build tables incrementally, you will want to limit the amount of data that you transform on each run of the pipeline for performance reasons. It can be inefficient to reprocess data that you know already exists in your target table and has been unchanged since the previous run of the pipeline. Without limiting data retrieval, you will inadvertently increase the number of times your database needs to scan both your source and target tables. In fact, building tables incrementally without limiting data retrieval will often be less performant than simply rebuilding the target table due to the increased number of scans. So far in this section, we have looked at examples of using dbt's incremental materialization where we didn't take this into consideration, but let's look at how we can solve this problem and improve the optimization of incremental models.

Fortunately, dbt has a couple of built-in methods that enable you to limit the data that will be transformed, processed, or scanned during subsequent invocations of models using the incremental materialization. The first of these features is the `is_incremental` macro, which enables you to write additional code in your model files that will only be included on subsequent runs of the model. A common pattern in incremental models is to add in a where clause that limits the data returned from the model's select statement based on the maximum timestamp found in the target table within the context of the `is_incremental` macro. Listing 4-10 shows an example of how we can use this pattern to extend the functionality of the incremental model, `fct_revenue_orders`, that we have been working with throughout this section.

Listing 4-10. Example showing how to use the is_incremental macro

```
{{
  config(
    materialized = 'incremental',
    unique_key = 'OrderId'
  )
}}
```

```
with get_orders_revenue as(
  select
    pro.OrderId,
    sum(pro.Price) as Revenue
  from {{ ref('int_orders_items_products_joined') }} as pro
  group by 1
)

select
  ord.OrderId,
  ord.OrderPlacedTimestamp,
  ord.UpdatedAt,
  ord.OrderStatus,
  ord.SalesPerson,
  rev.Revenue
from get_orders_revenue as rev
join {{ ref('stg_furniture_mart_orders') }} as ord
  on rev.OrderId = ord.OrderId
{% if is_incremental() %}
where ord.UpdatedAt > (select max(UpdatedAt) from {{ this }})
{% endif %}
```

The example in Listing 4-10 is an extension of Listing 4-7 from earlier in the chapter, and everything about the example is the same except for the final three lines of code. As seen earlier, we are using the `is_incremental` macro by evaluating the value that the macro returns using a Jinja if statement. We can use an if statement to tell dbt when to include the where statement in the compiled SQL because the `is_incremental` macro returns a boolean value and evaluates to true if these conditions are met:

- The target table exists in your database.
- The model materialization is configured to incremental.
- dbt is not running a full refresh of the model.

Note When working with incremental models, you may run into a scenario where your data has gotten out of sync. This can happen for many reasons such as a change in the logic or the introduction of a bug. Fortunately, dbt has a built-in flag that you can pass into dbt commands that allow you to rebuild incremental models from scratch. If you find yourself needing to do this, simply select the model like so: `dbt run --select my_incremental_model --full-refresh`.

While the pattern of using the `is_incremental` macro to dynamically include, or exclude, a where clause to filter data by a timestamp is the most common, it isn't the only way that you can use this macro. As a matter of fact, you can put absolutely any valid SQL within the context of it that you want to, so definitely experiment with it if there is some part of your transformation that you only wish to run on subsequent runs of your incremental models. For example, this can be especially helpful when building incremental models that contain window or aggregate functions.

Definition In this example, we use the {{ this }} function, which at compile time will be replaced with the target table's fully qualified name. We will discuss this function in much greater detail in Chapter 6. For now, just know that it is a handy way to reference your target table without hardcoding the name into your model definition.

Using the `is_incremental` macro to limit the data returned from the select statement in an incremental model can help improve the performance because potentially less upstream data gets scanned by the database. However, this is only one of the features that dbt offers to assist with limiting data retrieval in incremental models. There is a second feature that can be used to help improve performance of incremental models and is done by limiting the data that gets scanned in the target table. This is implemented by the `incremental_predicates` configuration, which accepts a string value that represents a valid string of SQL which will be injected into the SQL that dbt generates for you to insert, and update, data in the target table of the incremental model.

Caution The `incremental_predicates` configuration only works when using the merge strategy.

In Listing 4-11, we further iterate on the example that we have been working on this section by utilizing the `incremental_predicates` configuration. Let's first see what it looks like to use this configuration and then discuss later.

Listing 4-11. Example of using the incremental_predicates configuration

```
{{
  config(
    materialized = 'incremental',
    unique_key = 'OrderId',
    incremental_predicates=['DBT_INTERNAL_DEST.UpdatedAt > dateadd(day, -7,
    current_date())']
  )
}}

with get_orders_revenue as(
  select
    pro.OrderId,
    sum(pro.Price) as Revenue
  from {{ ref('int_orders_items_products_joined') }} as pro
  group by 1
)
...
```

In Listing 4-11, we set this configuration with this SQL:

```
'DBT_INTERNAL_DEST.UpdatedAt > dateadd(day, -7, current_date())'
```

When dbt runs the merge statement to incrementally load data into the destination table, it will inject this predicate into it as a condition that has to be met. What this statement is doing is limiting the data from the target table by only considering data for merge if it has been updated in the past seven. This can help performance because your database optimizer will not need to scan every record or partition in the target table to determine if a record should qualify for the merge. Listing 4-12 shows what the merge statement will look like with this predicate added in.

Listing 4-12. dbt-generated merge statement with a defined incremental_predicate injected into it

```
merge into dbt_learning.public_finance.fct_revenue_orders as
DBT_INTERNAL_DEST
    using dbt_learning.public_finance.fct_revenue_orders__dbt_tmp as
    DBT_INTERNAL_SOURCE
on (DBT_INTERNAL_DEST.UpdatedAt > dateadd(day, -7, current_date()))
and (
DBT_INTERNAL_SOURCE.OrderId = DBT_INTERNAL_DEST.OrderId
)
when matched then update set
...
when not matched then insert
...
;
```

Notice that in the predicate we defined, we used the alias `DBT_INTERNAL_DEST` before the name of the `UpdatedAt` column. This is done so that the database knows which column we are actually wanting the predicate to apply to because this is a merge statement and `UpdatedAt` would be an ambiguous column name without prefixing with an alias. As such, you should use `DBT_INTERNAL_DEST` and `DBT_INTERNAL_SOURCE` to prefix the columns in your predicates with respect to you want your predicate to function.

In this example, we use the `DBT_INTERNAL_DEST` prefix because we only want to consider data for this merge when the `UpdatedAt` column in the destination table is within the last seven days. While the data we are working with for this example is small, running a similar operation could produce valuable performance gains and a reduction of scans of the destination table. With that said, keep in mind that dbt won't validate the logic, or validity, of your SQL that you define in the `incremental_predicates` configuration, and as such this should be considered an advanced usage of dbt. For that reason, we recommend you only use this feature if you are working with large volumes of data and the benefits outweigh the complexity of implementation. As we've mentioned before, keep it simple for as long as you can!

Incremental Strategies

So far in this section, we have discussed how incremental models work from the standpoint of using the merge strategy. We took this approach because it is the default for Snowflake, which is a very popular data platform to pair with dbt and the platform that we built our examples on. However, we fully respect that not everyone will be using Snowflake, and the merge strategy doesn't fit everyone's use case for incrementally loading data. This is also recognized by dbt, and to solve this there are four strategies across the dbt ecosystem that can be used to configure the behavior of incremental models. Table 4-2 shows each of these strategies, how they work, and how complex they are to implement.

Table 4-2. *dbt incremental strategies*

Incremental Strategy	Complexity	Use Case
Merge	Low	Data needs to be inserted and updated in destination tables, but not deleted. Uniqueness is important to you
Append	Low	Data only needs to be inserted into destination tables. Uniqueness is not important to you
Delete+Insert	Medium	Data needs to be inserted and updated in destination tables, but not deleted. Uniqueness is important to you
Insert Overwrite	High	Data needs to be inserted and updated in destination tables. Uniqueness is important to you. Data must be stored in partitions because they are replaced/truncated on each run

Since we spent the earlier part of this section working through examples of using the merge incremental strategy, let's move on to look at how the others work. We will first show you how the delete+insert and append strategies are configured, but we will spend the majority of the remainder of this section walking you through an example of using the insert overwrite strategy.

Changes that must be made to the config block to use the delete+insert are very straightforward because all that you need to do is set the `incremental_strategy` configuration. Otherwise, the config block and the remainder of the model will stay unchanged. Listing 4-13 shows the syntax for configuring the delete+insert strategy.

Listing 4-13. Syntax for configuring an incremental model to use the delete+insert strategy

```
{{
  config(
    materialized = 'incremental',
    unique_key = 'OrderId',
    incremental_strategy = 'delete+insert'
  )
}}
...
```

Notice that we still define a unique_key configuration for the delete+insert strategy because dbt needs to understand the grain of this table so that it can appropriately delete or insert records to the destination table. This strategy works by deleting records in your target table when they are included in your transformation and subsequently reinserting them. In our opinion, this is easier to understand as a visualization. In Figure 4-1, you can see how this strategy works. In this image, we show a source table with three records, but the target table only contains two currently. On an incremental run of this model, records two and three are transformed. As a result, record two is deleted from the target table and then reinserted along with record three. The final result is a target table with all three records.

Figure 4-1. *The flow of data in the delete+insert incremental strategy*

The next strategy is the append strategy, and if you review Listing 4-14, you will notice that when using the append strategy we haven't defined a unique_key. dbt doesn't need to be aware of the grain of the destination table when using the append strategy because data will be incrementally loaded using an insert statement. As such, neither dbt or your database will check to see if duplicate records exist in this table. We recommend avoiding this strategy if there is any chance that duplicate data could enter your target table. Additionally, no existing data will be deleted or updated, so this is an important note to keep in mind when determining which incremental strategy is right for your model.

Listing 4-14. Syntax for configuring an incremental model to use the append strategy

```
{{
  config(
    materialized = 'incremental',
    incremental_strategy = 'append'
  )
}}
...
```

The final incremental strategy to cover is the `insert_overwrite` strategy, which is available for adapters such as BigQuery and Spark. This strategy is the most complex to implement, but can have the greatest performance benefit if set up correctly. This strategy is most similar to the merge strategy, but differs because instead of inserting and updating records in the destination table, it replaces and creates entire partitions of the destination table. Since this strategy makes use of partitions in this way, the number of scans that are required when incrementally loading this table is significantly reduced. This happens because only the partitions which you've instructed dbt to interact with are scanned, and as such performance improvements can be seen when working with very large volumes of data. If it isn't obvious, this strategy will only work if you have set up your target table to be partitioned on some column's values. It is common to use a timestamp or date column to partition tables.

Tip If you aren't working with large volumes of data, but still want to use the incremental materialization, we recommend that you start with the `merge` strategy and only move to `insert_overwrite` when the implementation complexity is justified relative to the performance gains.

When using the `insert_overwrite` strategy, there are two ways to configure which partitions will be overwritten:

- **Static partitions**: You tell dbt which partitions to replace.
- **Dynamic partitions**: You let dbt determine which partitions to replace.

For our first example of using the `insert_overwrite` strategy, we will configure the model to use the dynamic partitions method. In our opinion, this is the best place to start learning about this incremental strategy because it is the easier of the two methods to implement. You can find this example in the book's repository at this path: *~/models/marts/finance/fct_revenue_orders_insert_overwrite_dynamic.sql*. It is worth noting that we have disabled this model by using the `enabled` configuration because it will only work with certain adapters, but we will discuss this and many other model configurations later in the chapter.

Note While most examples in this book use Snowflake syntax, both examples as follows use BigQuery syntax because Snowflake doesn't support the `insert_overwrite` strategy.

The code in Listing 4-15 should look familiar to you as it's another alteration of the incremental model that we have worked with throughout this section, but there are a few distinctions in how the code for this variation of the model has been written.

Listing 4-15. Example of an incremental model using the insert_overwrite strategy and the dynamic partitions method

```
{{
  config(
    materialized = 'incremental',
    partition_by = {
        'field': 'UpdatedAt',
        'data_type': 'timestamp'
    },
    incremental_strategy = 'insert_overwrite',
    enabled = false
  )
}}
```

```
with get_orders_revenue as(
  select
    pro.OrderId,
    sum(pro.Price) as Revenue
  from {{ ref('int_orders_items_products_joined') }} as pro
  group by 1
)

select
  ord.OrderId,
  ord.OrderPlacedTimestamp,
  ord.UpdatedAt,
  ord.OrderStatus,
  ord.SalesPerson,
  rev.Revenue
from get_orders_revenue as rev
join {{ ref('stg_furniture_mart_orders') }} as ord
  on rev.OrderId = ord.OrderId
{% if is_incremental() %}
where date(ord.UpdatedAt) >= date_sub(date(_dbt_max_partition),
interval 2 day)
{% endif %}
```

Let's start by reviewing the config block:

```
{{
  config(
    materialized = 'incremental',
    partition_by = {
        'field': 'UpdatedAt',
        'data_type': 'timestamp'
    },
    incremental_strategy = 'insert_overwrite'
  )
}}
```

The config block from Listing 4-15 looks very similar to other examples that we have seen throughout this section. Of course, the incremental_strategy has been set to insert_overwrite, but also notice the inclusion of the partition_by configuration. As mentioned earlier, this strategy works by replacing entire partitions, so dbt will build your tables with the partition definition within this configuration. The partition_by configuration accepts a dictionary, and within we have defined which field to partition by and the data type of that field. In other words, when dbt builds this table, it will be partitioned on the UpdatedAt timestamp. Partitioning tables can be a complex subject, so we will refer you to your data platform's documentation to determine how your tables should be partitioned. Overall, by including the partition_by dictionary and setting the strategy to insert_overwrite, your model is now set up to utilize this strategy.

```
{% if is_incremental() %}
where date(ord.UpdatedAt) >= date_sub(date(_dbt_max_partition),
interval 2 day)
{% endif %}
```

As we discussed earlier, this example is using the dynamic partitions strategy, which means that we let dbt determine which partitions should be replaced. This is done by utilizing the _dbt_max_partition variable, as seen earlier. When the model gets run, this variable will be replaced with a timestamp corresponding to the maximum partition in the destination table. Furthermore, the timestamp is then transformed slightly by using the date_sub function so that we can effectively insert or overwrite partitions for the date of the maximum partition in the table minus two more days.

But don't think that you are tied to this logic, instead know that the incremental logic is fully configurable. For example, you may run this transformation much more frequently, and you only need to create or replace partitions for the past three hours. It's left up to you to understand your data needs to determine which partitions you should be creating or replacing.

Tip The is_incremental macro works the same for the insert_overwrite strategy as it does for all other incremental strategies.

In the prior example, we chose to use the dynamic method of running the insert_overwrite strategy, where we let dbt determine which partitions to replace based on the maximum partition in the destination table. While the dynamic method is simple to implement, it does come with some performance implications when working with large data volume because it includes an extra step where dbt needs to query the destination table to determine the maximum partition. However, this limitation can be overcome by making use of the static method. An example of using this method can be seen in Listing 4-16, or found in the repo here: ~/models/marts/finance/fct_revenue_orders_insert_overwrite_static.sql.

Listing 4-16. Example of an incremental model using the insert_overwrite strategy and the static partitions method

```
{% set partitions_to_replace = [
    'timestamp(UpdatedAt)',
    'timestamp(date_sub(UpdatedAt, interval 2 day))'
   ]
%}

{{
  config(
    materialized = 'incremental',
    partition_by = {
        'field': 'UpdatedAt',
        'data_type': 'timestamp'
    },
    incremental_strategy = 'insert_overwrite',
    partitions = partitions_to_replace
  )
}}

with get_orders_revenue as(
...

{% if is_incremental() %}
where timestamp_trunc(ord.UpdatedAt, day) in ({{ partitions_to_replace |
join(',') }})
{% endif %}
```

The first difference in this example, as seen in the following code, is that a Jinja variable has been defined at the top of the model. This variable serves to store a list of the partitions which we intend to replace when this model is run. In this example, we have defined that we want to replace the partitions for the past three days (today, yesterday, and the day before yesterday). This differs from the dynamic method because the maximum partition in the destination table is not taken into consideration when determining which partitions to replace.

```
{% set partitions_to_replace = [
    'timestamp(UpdatedAt)',
    'timestamp(date_sub(UpdatedAt, interval 2 day))'
  ]
%}
```

The other notable difference between the two methods for determining which partitions to replace is the block that gets included on incremental runs of the model. Notice in the following code that instead of using the `_dbt_max_partition` variable, we create a where clause that checks to see if the date of the record is in one of the dates in the list that we defined (`partitions_to_replace`):

```
{% if is_incremental() %}
where timestamp_trunc(ord.UpdatedAt, day) in ({{ partitions_to_replace |
join(',') }})
{% endif %}
```

So that it is put into the correct SQL syntax, we use the `join()` function so that the dates are in a comma-delimited list. For example, if today was 2023-01-03, then the where clause would compile to

```
where timestamp_trunc(ord.UpdatedAt, day) in ('2023-01-03', '2023-01-02',
'2023-01-01')
```

On a final note, we recommend that you only move to the static partitions method when the dynamic partitions method is not performant enough for you. Additionally, on a more general note about the incremental materialization, this is a powerful tool that you can use as part of dbt, but we recommend you proceed with caution. Incrementally building tables adds additional complexity to your data pipelines, so if you are a small team or working with small volumes of data, then the additional complexity may not be worth it. If this is the case, we recommend that you defer back to the table and view materializations for your data transformations.

Ephemeral

The final SQL model materialization that is available today in dbt is the ephemeral materialization. Ephemeral models work by allowing you to write SQL, store it in a .sql file, and then dbt will inject this SQL into any models that reference it by using a CTE (common table expression). In other words, dbt doesn't actually materialize ephemeral models in your database, but instead uses them as a way to allow you to modularize repeatable code and expose it for reference within other downstream models.

Reminder Models using the ephemeral materialization will not persist as objects in your database!

Let's quickly take a look at an example of how this works and then discuss when we think it makes sense to use the ephemeral materialization instead of something else. In Listing 4-17, you can see an example of how an ephemeral model is defined and then subsequently used in a downstream model.

Listing 4-17. Example of defining an ephemeral model and using it in a downstream table model

```
{{
  config(
    materialized = 'ephemeral'
  )
}}

select
  *
from {{ source('source_a', 'table_a') }}
```

```
/*************/
#Above, is an ephemeral model selecting from a source.
#Below, is a table model that includes the ephemeral model as a reference.
/*************/

{{
  config(
    materialized = 'table'
  )
}}

select
  *
from {{ ref('my_view_model') }} as my_view_model
join {{ ref('my_ephemeral_model') }} as my_ephemeral_model
  on my_view_model.id = my_ephemeral_model.id
```

As you can see in Listing 4-17, ephemeral models are defined in .sql files just like all other SQL models and can be referenced downstream using the `ref` function. The model file looks the same as any other model file would, but if you take a look at Listing 4-18, you will see how dbt injects the SQL from the ephemeral model into the compiled SQL of the downstream model.

Listing 4-18. Compiled SQL of the table model defined in the second part of Listing 4-17

```
with __dbt__cte__my_ephemeral_model as(
  select
    *
  from source_database.source_a.table_a
)

select
  *
from dbt_learning.public.my_view_model as my_view_model
join __dbt__cte__my_ephemeral_model as my_ephemeral_model
  on my_view_model.id = my_ephemeral_model.id
```

In the second half of the example, the join that referenced the ephemeral model was replaced with a join that references the CTE. This is a nice feature to be aware of, but in our experience there are often better materializations to use than the ephemeral materialization.

In our opinion, the only time that ephemeral models are useful is when you have a model that is starting to have a lot of code stored in the same file. At that point, you may consider pulling some of your logic up a layer into one or two ephemeral models, but be aware that this is only for readability purposes. It is unlikely that ephemeral models will provide you any sort of performance improvement. Furthermore, if you are referencing ephemeral models downstream in more than one or two models, we would recommend that you consider converting the model to use the view or table materialization.

Python Models

From the very beginning, dbt was built to be a framework to introduce software development best practices into the world of analytics. For a long time, this meant that you could use dbt to build models using SQL, and strategically this made sense as SQL has long been the lingua franca of data professionals. However, those early days of dbt are gone, and the tool has proven to be useful for a wide range of use cases. While SQL can often be the best tool for a data transformation, this isn't always the case. So, by popular demand, dbt started supporting Python models for the three major players in the cloud data platform landscape, including Snowflake, BigQuery, and Databricks.

By using Python models, you are able to write data transformations that are expressed using syntax that is familiar to many data engineers and other data professions. For both BigQuery and Databricks, Python models are written using the PySpark framework, but Snowflake is slightly different and uses its own framework called Snowpark. Although the two frameworks are different, they share many similarities and allow us to talk about Python models fundamentally in the same way. Regardless of which data platform you are using, Python models in dbt are represented as a function that operates on and returns a dataframe object. If you have used Pandas, PySpark, or Polars before, then you are already familiar with the concept of a dataframe. However, if you are new to dataframes, then in their simplest form you can just think of them as a table-like object or a representation of rows and columns of data.

Also, similarly to SQL models, dbt will not execute your Python models on its infrastructure, but instead will compile the code correctly and ship it off to your data platform to be executed. With that said, since Python models will be executed on your data platform's infrastructure, it is up to you to ensure that you have properly set up your data platform to support accepting Python jobs. This may range from simply enabling Python capabilities to defining Python-specific compute clusters. At the time of writing this book, Python models are still a very new feature, so we will refer you directly to dbt's documentation for further instructions on ensuring you are set up to run Python models: `https://docs.getdbt.com/docs/build/python-models#specific-data-platforms`.

Writing a Python Model

While there are many vast differences between how SQL models and Python models are written, there are a few similar concepts that span across the two different model types. First, Python models are stored within the models directory just the same as SQL models. This is useful because you can still keep similar models stored in the same directories regardless of whether they are SQL or Python models. It would be quite a struggle to need to store Python models in a fundamentally different way.

Other similarities include referencing upstream models and sources, setting model-level configurations, and being able to then reference Python models in other downstream models (whether those being Python or SQL doesn't matter). Take a look at Listing 4-19 to see how a very simple Python model can be defined.

Listing 4-19. Example of the basic Python model syntax

```
def model(dbt, session):
    dbt.config(materialized='table')
    raw_orders_df = dbt.ref('raw_products')

    final_df = raw_orders_df.WithColumn(
        'close_out_price',
        raw_orders_df.price * 0.6
    )

    return final_df
```

To break this example down, let's start with line 1:

```
def model(dbt, session):
```

Whenever you write a Python model, the function that returns the dataframe that gets materialized as a table in your database must always be named `model`. Following that, this function will always accept two positional arguments. These arguments are used for

- **dbt**: A Python class from dbt Core that enables this function to be exposed in your dbt project as a model.
- **Session**: A Python class used to connect to your data platform. This is how data is read into dataframes within a model function and how dataframes that get returned are written to tables in your database.

```
dbt.config(materialized='table')
```

Next, notice that directly within the function, we access the `config` method of the `dbt` class. Here, we are defining that the dataframe returned from the model should be materialized as a table in the database. This config method works similarly to the config block in SQL models, but is currently more limited. For a full list of the most up-to-date configs that can be set in Python models, we recommend you check out the dbt documentation.

Additionally, note that we configured the model to use the table materialization. If you recall from earlier in the chapter, SQL models have several materializations available for you to use with them. This is still true with Python models, but the scope of materializations is more limited. In fact, you can only configure Python models using the table or incremental materialization. Though, thinking about this makes sense because Python models return a real dataset, and as such it wouldn't make sense for you to be able to configure them to use the view or ephemeral materializations. Furthermore, these two materialization types are inherently SQL based and don't make sense in the context of a Python model.

```
raw_orders_df = dbt.ref('raw_products')
```

Now that the function has been created and configured, we need data to transform. We can get data from upstream models by using the `ref` method of the `dbt` class. Here, we get the data from the upstream model *raw_products* and store it in a dataframe. Since we're referencing an upstream model, we used the `ref` method, but the `dbt` class also has a `source` method that you can use if you need data directly from a source table.

For the remainder of this example, we conduct a simple transformation by adding a new column which represents the closeout price of a product where the closeout price is a 40% discount. We then store the entire resulting dataframe in a new dataframe named `final_df`. This dataframe is what gets returned from the function when dbt is invoked, and the result gets stored as a table in the database. Just as we've done in this example, for your Python models to materialize correctly they should always return a single dataframe object.

We have intentionally made this example simple because we only want to cover the key pieces of defining Python models such as the function naming convention, expected positional arguments, and expected return values. There are simply too many syntactic differences in how to actually write Python transformations depending on what data platform you are using. However, we do encourage you to explore using Python models as they can be a powerful addition to your standard SQL models.

When to Use Python Models

While we wish it were easy to draw a line in the sand and say, "Use Python for X and SQL for Y," it unfortunately is not that simple today. This is mainly true because at the time of this writing, Python models are still an extremely new feature of dbt. As such, we don't have much firsthand experience in using many Python models in our projects. That isn't to say that they don't have a place because they certainly do.

For our standpoint, when trying to decide between using a SQL or Python model, we recommend that you default to using SQL models and only move to a Python model when your code would execute more performantly or if the development process would be more simple. For example, if you find yourself writing a SQL transformation that includes a significant amount of Jinja, such as a dynamic pivot, and you could instead utilize PySpark or Snowpark syntax to simplify that transformation, it may make sense to write the model as a Python model.

Lastly, we believe the most lucrative use case for Python models is opening the door more broadly to data science and machine learning teams. Before the addition of Python models, if one of these teams wanted to train a machine learning model using data that was the result of a dbt SQL model, then they would have needed to use some external tool to manage the infrastructure to run their Python code and possibly even another tool to orchestrate that process. But, with the addition of Python models, these teams can easily integrate into the same stack as the other data teams and use dbt to build and orchestrate their machine learning models.

Overall, the addition of Python to dbt was a huge step forward in expanding the capabilities of dbt and making the tool accessible to a broader audience. However, we encourage you to use Python models only when writing a SQL model will be too complex, or if SQL simply can't provide you the capabilities that you need for your transformation.

Modular Data Transformations

So far throughout the chapter, we have talked about how to build models, and in the prior chapter, we discussed how to define sources and seed files. We've also covered how models can reference other upstream sources, seeds, and models, and if you recall from Chapter 2, we discussed the topic of project structure. Within that section, we made recommendations on how you can organize the models in your project. Recall that there is more than one way to structure your dbt project, and no one way is right or wrong. With that said, many teams like to follow the staging-intermediate-marts pattern when it comes to building their models. In this pattern, you end up with three types of models:

- **Staging models**: These models are one-to-one with sources. They are meant to conduct very light transformations, such as casting data types and renaming columns. Staging models shouldn't contain any joins, and no other model type besides staging models should select from sources.
- **Intermediate models**: These models should select from upstream staging models and conduct transformations that are reusable by models downstream in your transformation pipeline.
- **Mart models**: These include both fact and dimension models and should represent the models which are exposed to end users in your data warehouse. These models should be the result of transformations of staging and/or intermediate models.

Building dbt models into these three categories lends itself well to the concept of modularity. Modularity is a concept, and best practice, in Software Engineering that says that code should be broken out into reusable chunks so that a set of logic is only implemented once and then reused. By writing your data transformations in logical steps using modularity, you gain several benefits, including

- Transformations are reusable in downstream models. No longer do you need to conduct the same join patterns over and over again.
- When failures happen, and they *will* happen, debugging is much more simple because dbt will tell you exactly which model caused the failure. If you have broken your transformations up to be modular, then you should be able to quickly identify the root cause.
- Modularity adds a layer of abstraction and can hide the complexity of data transformations.
- Most importantly, modularity improves the scalability of your data transformation. Do you need to add a new fact table? No problem because most likely the staging and intermediate models that you need are already there for you, so you can spend more time on delivering value to the business and less time worrying about how to get the data you need.

Now, that's not to say that you can't follow a pattern of monolithic data transformations and write all of your transformation code for a fact or dimension table in a single .sql file. That will work, but you should ask yourself if that is the most efficient way for you and your team to work. With that said, the staging-intermediate-marts pattern is not the only one that you can use to achieve modularity. You can also implement a pattern that doesn't create staging models that are one-to-one with source tables, and in fact we have seen many teams do this and be successful doing so. This pattern is somewhat less modular, but there are teams that instead use the staging layer to house all of their transformation logic, test transformations there, and then promote it to mart models once everything checks out. This is often referred to as the Write-Audit-Publish (WAP) pattern. We discussed this in Chapter 2, but it is worth mentioning again because there are so many different ways to structure your project, and we don't want you to think that you are pigeonholed into one way of doing things. Following this pattern removes the burden of needing to maintain an entire transformation layer that is simply a cleaned up copy of your source data. On a final note, these patterns aren't mutually exclusive, and most aren't. You really can dream up whatever pattern you believe lends itself best to your data, your requirements, and your use case. There are a lot of big opinions in the world of data, some are great and some aren't, so we encourage you to be intentional when trying to determine what will work best for you.

However, if you do choose to implement your models in a modular fashion, your DAG will be easier to understand, but we do caution you to not overabstract your project. There is a fine line between modularity for ease of use and modularity that makes things overly confusing. We can't tell you exactly where that line will get drawn because it heavily depends on the business logic that your use case requires, but we can caution you to consider this when choosing how modular you build your transformations.

We have mentioned the DAG before in previous chapters, but as a reminder, it is a core component of dbt that allows it to interpret how your different transformations are related to each other. The DAG (directed acyclic graph) is a lineage graph that shows how all of the objects in your dbt project are related to one another. While the DAG isn't a concept that is unique to dbt, it is vital to understand the importance of you being able to interpret the DAG that your project produces.

Tip You can produce the DAG for your project by running `dbt docs generate` and `dbt docs serve`. Or, if you are using dbt Cloud, you will be able to see it in the cloud IDE.

Suppose that we didn't write our transformation in a modular fashion, and we have one fact table and two dimension tables that have been modeled. In Figure 4-2, you can see an example of a DAG that shows data flowing directly from sources into these mart models.

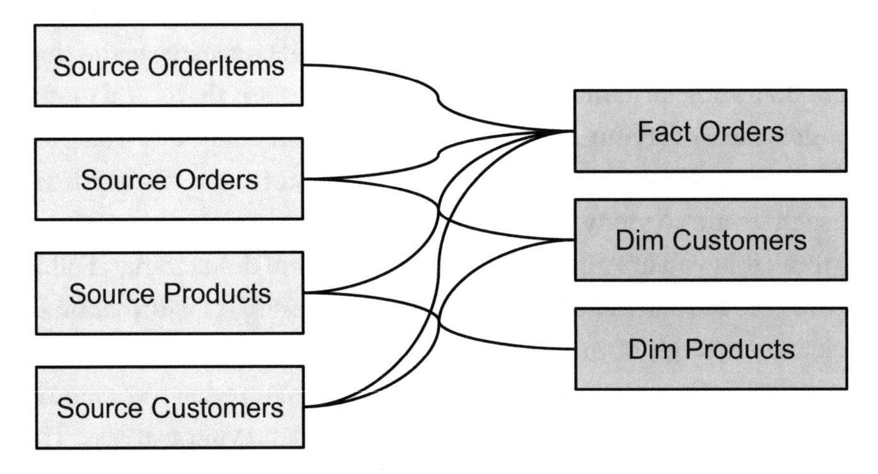

***Figure 4-2.** A messy DAG*

While this DAG is fairly easy to understand right now because we only have three mart models at the end of the DAG, in the real world it is common to have hundreds, or even thousands, of mart models. As you can imagine, if we were to query data directly from the source tables every time, then we would quickly end up with a very messy DAG that is hard for anyone to understand. However, we again recommend that you consider the benefit of abstraction vs. understandability. Overabstraction is just as bad, if not worse, than no abstraction at all. However, if you were following the staging-intermediate-marts pattern, a few changes could be made to introduce more modularity so that the DAG is much more scalable and easy to understand. Take a look at Figure 4-3 to see what this DAG would look like if we refactor it to be more modular.

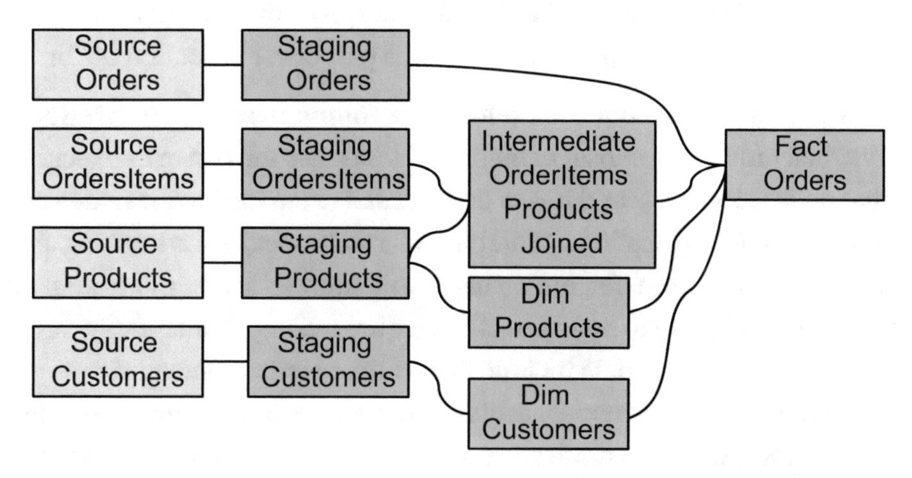

Figure 4-3. *A modular DAG*

As you can see by adding a staging layer between source tables and anything else downstream, we can start to clean the DAG up significantly. Then, we introduce an intermediate model to join order items and products together, and finally the fact orders table is created as a combination of other upstream models including staging, intermediate, and dimension models. While we didn't make many changes, we did add enough modularity to make this DAG more readable because there is now less overlapping flow of data.

Node Selection

We discussed running models during Chapter 1 and provided you with an extensive list of the command-line flags that can be included with your commands. However, now that we have provided you some examples of building models, you may be wondering how you can run your models. As a reminder, there are two commands that can be used to materialize your models as objects in your database:

- `dbt run`: This command will execute all of the models in your dbt project and materialize them as objects in your database.
- `dbt build`: This command will do the same as `dbt run`, but also include all seeds, snapshots, and tests as part of the execution of dbt.

While these commands alone are useful, it is a common pattern to only want to run an individual model or a subset of models. So, let's use the rest of this section to discuss what node selection is and how it can be used to customize the nodes that will be included in your dbt invocations. When we refer to "nodes," we are talking about most objects that you define within dbt, but most notably sources, seeds, models, snapshots, and tests. So, when referring to node selection, we are talking about modifying a dbt command, such as run or build, to include or exclude specified nodes.

Node selection in dbt is extremely flexible, but this flexibility comes with a learning curve as you start working with more advanced selectors. Since there are so many different ways to approach node selection, we will start with the basics and offer examples that grow in complexity. We won't cover every possible concept because there are so many, but by the end of this section, you should be comfortable with the syntax of node selection and be equipped with the knowledge necessary to build out fairly complex selectors in dbt. We will cover

- Node selection basics
- Graph and set operators
- Selection methods
- YAML selectors

To get started, let's look at the basics of node selection, but you may be surprised because you have already seen node selection syntax throughout a few examples in this book. In its most basic form, node selection starts with including a flag and a node name with the command that you are issuing to dbt. For example, suppose you have a model named `fct_daily_orders` and you want to issue the `dbt run` command against this model, but only against this model. To do this, we will use the `--select` command-line flag. The full command to run this operation looks like this:

```
dbt run --select fct_daily_orders
```

But, what if you wanted to do the inverse of this command and run every model in your project **except** for the fct_daily_orders model? This is where the --exclude command-line flag comes in. We can refactor the selection criteria to this to achieve this requirement:

```
dbt run --exclude fct_daily_orders
```

This is a good introduction to the concept of node selection, and running individual models is useful in a development environment, but realistically your requirements for node selection in a production job are likely to be much more complex. Maybe you need to run all of the nodes that are downstream from a certain node, only models with a specific tag, or only models that intersect between two distinct selectors. Fortunately, dbt offers the flexibility with node selection to achieve all of these use cases individually, and you could even group them all together. This brings us to the next, and slightly more complex, concept within node selection: graph and set operators.

Tip For most of this section, we will focus on using different selectors coupled with the --select flag, but you can use all of these in combination with the --exclude flag as well.

Graph operators are used to select nodes that share lineage in some fashion. Another way to think of this is by looking at your DAG and following the arrows between nodes. Graph operators enable you to select nodes for invocation based on how they are connected in the DAG (i.e., the graph). The most straightforward graph selector is the plus sign (+) selector. This allows you to select nodes that are either upstream or downstream from your selected node. For example, suppose you want to run the `fct_daily_orders` model and all of the models upstream of it. You could write this command as

```
dbt run --select +fct_daily_orders
```

It is worth noting that the plus sign selector is **fully inclusive**, meaning that in our preceding example the `fct_daily_orders` model will be run, its parent(s), its grandparent(s), its great grandparent(s), and so on until there is no more upstream lineage to run. As you may have guessed, you can move the plus sign selector to after the model name in the command as well if you want to run itself, its children, grandchildren, and so on. That would be written like this:

```
dbt run --select fct_daily_orders+
```

Maybe it isn't always necessary for you to run the entire upstream or downstream lineage of a model, and instead you want to limit the ancestry that is included in your node selection. Of course, dbt has a selector that you can use to modify the plus sign selector. You can do this by utilizing the n-plus selector. With this selector, you simply add a number before or after the plus sign depending on whether you are attempting to run upstream or downstream models. We have provided a few examples as follows to show how the n-plus selector works for upstream, downstream, and selection of both:

```
##Run fct_daily_orders and its parent models.
dbt run --select 1+fct_daily_orders

##Run fct_daily_orders, its children, grandchildren, and great
grandchildren.
dbt run --select fct_daily_orders+3

##Run fct_daily_orders, its parents, and children.
dbt run --select 1+fct_daily_orders+1
```

There are other graph selectors such as the at (@) and star (*) selectors. The @ operator can be used to select a model, its children, and its children's indirect parents (we recognize the parent-child relationship analogy starts to break down here). This is helpful in situations when the other parents of downstream models may not exist or are stale because this selector will ensure they are run. Here is an example of the syntax of this selector:

```
dbt run --select @fct_daily_orders
```

The final graph selector we will cover is the star (*) selector, and it is conceptually very simple. It is just used to run all of the nodes within a particular directory or package.

For example, if you wanted to build all models, and related nodes, within a file path of models/marts/product, then you would structure the command such as

```
dbt build --select marts.product.*
```

Notice a few things about this command:

- You don't need to include the models directory in the selector.
- Directories are separated using dot notation instead of your typical forward slash.
- This example uses the build command, which is completely valid for node selection. For the most part, you can use either run, test, build, or seed with regard to node selection.

Next up, let's take a look at set operators which can be a bit tricky to grasp at first, not because the concepts are complicated, but because the syntax takes a few passes to remember. There are two types of set operators that you can use when you are selecting nodes: union and intersection.

The concept of union is fairly straightforward because you just provide two, or more, selectors and dbt will invoke all of the nodes that you selected. We like to think of the union set operator as a way to shorthand multiple dbt commands. Let's take a look at an example. Suppose that you still want to run the `fct_daily_orders` model, and its downstream dependencies, that we've used in other examples, but you also want to run another model named `fct_daily_returns` and all of its downstream dependencies. Of course, you could write two different run commands to do this, as follows:

```
dbt run --select fct_daily_orders+
dbt run --select fct_daily_returns+
```

But, the union operator allows you to shorthand this into a single command such as

```
dbt run --select fct_daily_orders+ fct_daily_returns+
```

As you may notice, there is nothing other than whitespace between the two node selectors, and that is what we meant by funky syntax. The presence of whitespace between `fct_daily_orders+` and `fct_daily_returns+` is how dbt infers that you want to union these selectors. We will leave it up to you whether you use the union operator or write individual commands, but in our experience, it is more declarative and clear to just write two separate commands. The only difference is that they won't be run in parallel, one command will need to be completed before the second runs. This may, or may not, be an issue for you, but it is something you should consider.

The second set operator is the intersection selector, which is used to run the selected nodes and **only** their common nodes. The syntax to capture intersections is done by using a comma (,) between selectors. So, for example, if we wanted to run the intersection of the `+fct_daily_orders` selector and the `+fct_daily_returns` selector, then we would write the command like this:

```
dbt run --select +fct_daily_orders,+fct_daily_returns
```

This is better represented as a visual, so in Figure 4-4 you can identify which models would be run by this command by looking at the nodes in the DAG that have a check mark in them.

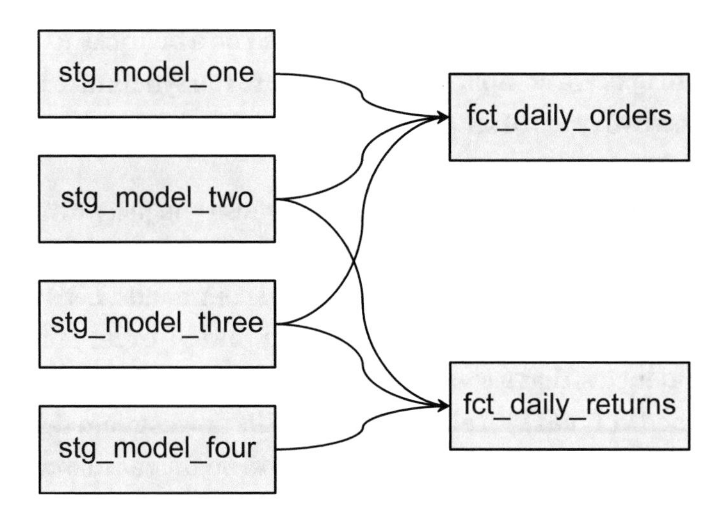

Figure 4-4. *Example of nodes run when using the intersection syntax*

So far in almost all of these examples, we have used model names to select which nodes to execute as part of the dbt command, but there are many other options at your disposal for selection. These are referred to as node selection methods, and there are many different ones that you can use. Since there are so many options, we aren't going to go into great detail on how each works. Instead, we have provided a list of methods and an example command corresponding to each. No matter which method you use, they will all follow the syntax of `method_name:selector`. If any of these seem interesting to you, we recommend that you explore creating commands using different combinations of these methods:

- **Tag**: `dbt run --select tag:finance`
- **Source**: `dbt run --select source:hubspot`
- **Resource type**: `dbt build --select resource_type:model`
- **Path**: `dbt run --select path:models/marts/product`
- **File**: `dbt run --select file:fct_daily_orders.sql`
- **Fqn**: `dbt run --select fqn:fct_daily_orders`
- **Package**: `dbt run --select package:dbt_artifacts`
- **Config**: `dbt run --select config.materialized:incremental`
- **Config (cont.)**: `dbt run --select config.schema:finance`
- **Test type**: `dbt test --select test_type:generic`
- **Test name**: `dbt test --select test_name:not_null`
- **State (more on this in Chapter 10)**: `dbt run --select state:modified`
- **Exposure**: `dbt run --select exposure:some_important_dashboard`
- **Result (rerun failed models)**: `dbt run --select result:error+ --state target/`
- **Source status**: `dbt source freshness && dbt run --select source_status:fresher+ --state target/`
- **Group**: `dbt run --select group:product`
- **Version**: `dbt run --select version:latest`
- **Access**: `dbt run --select access:private`

Last, but certainly not least, are YAML selectors, which are used by creating a YAML file that lists all of the selection criteria for a command. This provides a clean way to maintain production-grade selectors that are used in jobs. Additionally, it provides some structure to your selection criteria instead of it all being listed out in one long string. We recommend that you use YAML selectors when you have more than two types of selection criteria. This number is somewhat arbitrary, but we feel it is the sweet spot for when you should convert to YAML selectors, but your mileage may vary.

The following is the full syntax of YAML selectors:

```
selectors:
  - name: my_selector
    description: This is my YAML selector.
    definition:
      union | intersection:
        - method: some_method
          value: nightly
          children: true | false
          parents: true | false
          children_depth: 1
          parents_depth: 1
          childrens_parents: true | false
          indirect_selection: eager | cautious | buildable
        - exclude: same syntax as above...
```

Basic syntax is great, but let's take a look at a practical example. Suppose we want to select nodes that are intersected based off of three criteria:

- Has a tag of finance
- Is from the source of Quickbooks
- And is configured to be materialized incrementally

To achieve this, our YAML selector definition would look like this:

```
selectors:
  - name: incremental_finance
    description: This is used to run incremental finance models.
    definition:
      intersection:
        - method: tag
          value: finance
          parents: true
        - method: source
          value: quickbooks
```

```
      parents: true
    - method: config.materialized
      value: incremental
      parents: true
```

Now that this YAML selector has been built out, you can actually run it using a dbt command. So that tests, seeds, and snapshots are included, we could use the build command like this:

```
dbt build --selector incremental_finance
```

As you can see, node selection can be very simple, but can also progress to become very complex when you start using features such as set operators and YAML selectors. As with most things, we recommend that you start simple and gradually increase complexity when additional complexity has a benefit such as easier development or maintenance.

Model Configurations

To close out the chapter, we wanted to provide you some information about additional configurations that you can apply to your models. Throughout this chapter, we have talked about setting configurations at the model level using the config block, or config method for Python models, but this isn't the only way to set model configurations. In fact, there are three different places that make up the hierarchy of where model configurations can be set, which can be seen in Figure 4-5.

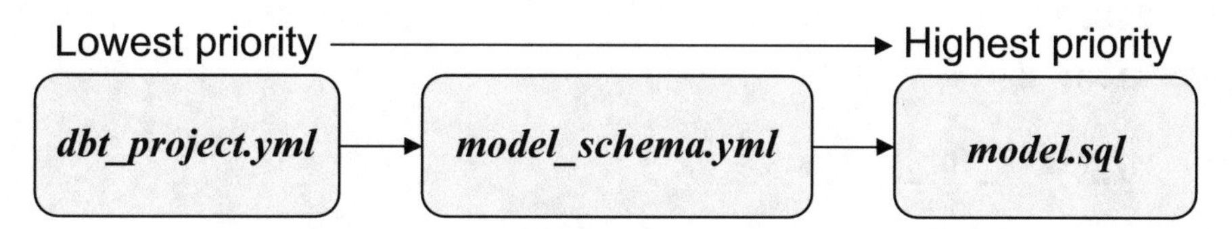

Figure 4-5. *The configuration hierarchy*

What this is showing is that dbt will use the configs at the lowest level of configuration. For example, if you define a materialization type in the dbt_project.yml and within your model's .sql file, then the configuration set in the .sql file will take precedence.

At this point, we have provided several examples of how to set model configurations directly within model files, but let's look at how to configure models within .yml files. In Listing 4-20, we have provided an example of how a model_schema.yml file can be organized, and then following that in Listing 4-21, we provide an example of how a dbt_project.yml file can be organized.

Listing 4-20. Example of a schema.yml file

```
version: 2

models:
  - name: stg_furniture_mart_products
    config:
      materialized: view
      tags: ['staging']
    columns:
      - name: ProductID
      - name: Product
      - name: Price
...
  - name: stg_furniture_mart_orders
...
```

Listing 4-21. Example of a dbt_project.yml file

```
name: 'my_first_dbt_project'
version: '1.0.0'
config-version: 2

profile: 'my_first_dbt_project'

model-paths: ["models"]
...
models:
  my_first_dbt_project:
    staging:
      +materialized: table
    intermediate:
```

```
      +materialized: view
    marts:
      +materialized: view
...
```

As you can see between these two examples, we have defined the materialization type of a model or set of models in different ways. For this example, we are using the materialized configuration, but keep in mind that most configurations can be set in different spots throughout your dbt project. Let's first look at Listing 4-20, the *model_schema.yml* file, to see how configurations are set. Within this file, we have named models individually which allows us to provide configurations, and for the stg_furniture_mart_products model, we have set the materialization to be a view. Also, within this file is where you will handle documentation and testing of your models, which we will discuss later in the book.

Tip The *model_schema.yml* can be named whatever you want; we just follow this naming convention. dbt will be aware of the purpose of the file simply by it being a YAML file in the correct format.

However, in Listing 4-21, we set the materialized configuration, but notice that it isn't being set for individual models. Instead, configurations within the *dbt_project.yml* apply to entire directories. We know that stg_furniture_mart_products lives within the staging directory, so according to the *dbt_project.yml*, this model should be materialized as a table. Well, if you recall the hierarchy in which configurations are applied, we understand that the stg_furniture_mart_products model will be materialized as a view because

- The materialized configuration was not set in the model file.
- So, dbt defers to the *model_schema.yml* file, where we set the configuration to equal *view*.
- Since the configuration was set in *model_schema.yml*, dbt will entirely ignore the configuration set in the *dbt_project.yml*.

By this point, you may be wondering what other configurations are available, and there are a lot of them! However, be aware that there are general configurations that work with all dbt adapters, and there are adapter-specific configurations that are tailored to work with your data platform. For the sake of simplicity, we are only going to cover the general configurations, but we encourage you to research the dbt documentation to understand what additional configurations are available for the adapter that you are using:

- **alias**: Overrides the naming of a relation.
- **database**: Specifies the target database for a model or seed.
- **enabled**: By setting to **false**, models, seeds, snapshots, and tests can be disabled.
- **full_refresh**: Sets a model to always, or never, fully refresh. A full refresh is when an incremental model is rebuilt from scratch.
- **contract**: Enforces that your model exactly matches the name, data type, and other attributes defined within YAML files.
- **grants**: Specifies grants that should be applied to models, seeds, and snapshots at runtime.
- **group**: Enables restricting access to reference private models from other downstream models.
- **docs**: Hides the node from dbt-generated documentation. Note that the node will still appear in your project's DAG.
- **persist_docs**: Persists documentation defined in YAML as column and relation comments in your database.
- **pre-hook**: Runs one or more ad hoc SQL statements before a model is run.
- **post-hook**: Runs one or more ad hoc SQL statements after a model is run.
- **schema**: Specifies the target schema for a model or seed. Useful if you don't want to build all of your models in your default target schema.
- **tags**: Creates one or more tags on nodes. Useful for grouping nodes and can also be used as a selector in CLI commands.

- **meta**: Creates a dictionary of arbitrary metadata that you want to be stored in the *manifest.json* that is generated by dbt. This metadata is visible in the documentation site.

Summary

Throughout this chapter, we introduced you to the most fundamental piece of dbt, models. Without models, the rest of dbt is nearly worthless, but as you will see in later chapters, much can be built off of models:

- Other models
- Tests
- Documentation

We started this discussion by covering SQL models, which for a long time were the only type of model you could build with dbt. In this chapter, we covered the different types of materializations that you can use with SQL models. A materialization defines how dbt will, or won't, persist the model in your database. There are four materializations that can be used for SQL models:

- **View**: Persists a view in your database. Often used as the preferred materialization type for staging models or other lightweight transformations.
- **Table**: Persists the data returned by your model in a table in your database. Often used in mart models (facts and dimensions) or other cases when query results need to be returned quickly.
- **Incremental**: Serves the same purpose as the table materialization, but this allows you to load the table incrementally instead of rebuilding each time dbt is invoked. Most useful when the table materialization doesn't provide the performance that you require.
- **Ephemeral**: A unique materialization that doesn't persist in your database. Instead, this returns a CTE that can be referenced in downstream dbt models.

Following our discussion on SQL models, we talked about the new kid on the block: Python models. As of the writing of this book, Python models are still in their early stage, so we didn't go into great detail on them. But, we did discuss the basics including how they fundamentally work, how they persist data as tables in your database, and how to use methods of the dbt class that is passed to Python models. Additionally, we briefly discussed when it makes the most sense to use these models. It's our opinion that models should be built using SQL if they can, but Python does open the door for additional complex workflows, including data science and machine learning. These data professionals used to have to work outside of dbt, but now can work directly in dbt when it makes sense to.

While it's important to understand how to build models, we have seen dbt projects quickly become an unmaintainable mess because of the lack of preparation and inadequate use of modularity. By writing your transformations in a modular fashion, your project becomes more maintainable because it is easier to reuse common transformations, more simple to debug failures, and it simply makes your DAG more easy to digest. We encourage you to follow the pattern of building staging, intermediate, and mart models so that your transformations will always be modular.

In the next chapter, we will discuss a different resource type that you can use in dbt: snapshots. Snapshots are a way for you to generate type 2 dimensions for tracking historical changes to source data when you have no better way to do it. We will provide examples of building snapshots, when, and why you may need to use them.

CHAPTER 5

Snapshots

It is likely that at some point during your time working with data, you have come across a dataset that changes, but the history isn't retained. This can become a nightmare to deal with because quite often data professionals need to query the data from a point in time. During this chapter, we will talk about the ways that dbt can help track row changes and maintain change history using snapshots. If you are familiar with dimensional data modeling, you may be familiar with slowly changing dimensions (SCDs), which allow you to track historical data in various ways (or Types). Snapshots are identical to a Type-2 SCD where a record is stored for both the current and changed rows in the same table and differentiated based on status and timestamp columns. In other data model designs, the concept exists, but may be referenced as a different name.

Consider a scenario where you have tables in your data warehouse that come from your company's application database, and in those tables data is frequently updated, but the history is not maintained. We can describe this data as being mutable, which is great for software developers, but less than ideal for data engineers, data analysts, and analytics engineers. Mutable data is ideal for software developers because they typically are only concerned with the current state of the data. However, when querying data that can change from the viewpoint of data analytics, having a record of the changes for a single row is valuable and can unlock insights that we may miss if we only have the current state stored. Primarily in transactional systems (though sometimes in analytical systems), teams will need to track changes. Whether for auditing purposes, troubleshooting, or replication reasons, teams will utilize change capture technologies. Most mainstream database technologies have change capture built into them and often track changes in system tables that can then be read to determine what has changed. In terms of SCDs, these work similarly to a Type-4 where historical changes/records are stored in a separate table. While not exactly the same, utilizing change capture can produce similar tracking to what you do with snapshots in dbt.

C. Cyr and D. Dorsey, *Unlocking dbt*, https://doi.org/10.1007/978-1-4842-9703-2_5

Take a look at Tables 5-1 and 5-2 to see an example of a customer table where the data changes, but the history is not maintained. In these examples, we see that the customer's state of residence changed from *New Hampshire* to *Florida*.

Table 5-1. *Example of a customer that lives in New Hampshire*

CustomerId	State	UpdatedAt
100	New Hampshire	2021-01-30 12:00:00

Table 5-2. *Customer 100 moves from New Hampshire to Florida*

CustomerId	State	UpdatedAt
100	Florida	2023-01-31 05:30:00

With the data stored in this format, we are able to answer questions with the data only regarding the present state, but what if the data was in a format similar to Table 5-3? Having the data stored in a format similar to this can enable you to unlock answers to questions such as how long this customer lived in New Hampshire or where they were living when they placed an order. Of course, this concept goes far beyond tracking the history of a customer and can be applied to virtually any mutable data source. Snapshots are dbt's answer to how we can track these changes, and throughout this chapter, we will discuss when you should use snapshots, how they work, how to configure them, and more.

Table 5-3. *Example of a customer with the history being tracked*

CustomerId	State	UpdatedAt
100	New Hampshire	2021-01-30 12:00:00
100	Florida	2023-01-31 05:30:00

When to Use Snapshots

During the introduction for this chapter, we discussed mutable data sources and snapshots at a very high level, and later in the chapter, we will discuss how you can implement snapshots and the technical details surrounding doing so. But, let's first dig deeper into when it is best to use, and avoid, the snapshot feature of dbt during development of your data warehouse.

Snapshots have one primary use case, and that can be summarized as enabling you to track changes to data sources that don't store history. This is a very simplified version of what a snapshot does, but it gets the point across. An additional point to call out is that it is almost always recommended to snapshot sources and then treat the snapshot as a source in downstream modeling, meaning you should typically `select * from {{ some_source }}` to simply track the changes to the source data. You could implement transformations within a snapshot, but it is not recommended because snapshots by nature are not idempotent. As such, if you were to need to rerun your transformation, you can't guarantee that the resulting output will be the same every time.

Note Idempotent, or idempotence, is a term often used in Data Engineering to describe a pipeline or transformation that given a set of inputs will always yield the same output.

To solidify your understanding of the concept of snapshots, let's take a look at a couple of examples of when you could use snapshots.

Example One

For this example, let's consider the *orders* data that is included in the book's GitHub repo. This is a great example of a mutable data source because this represents data that changes and could do so quite frequently in practice. Notice that the orders in the dataset fall into one of three statuses: placed, shipped, and returned. As you can imagine, it would have been lucrative to have maintained the history for each order as the status, or other fields, had been updated. This is where dbt snapshots could come in to help out with maintaining a history of row changes.

If you want to follow along with this example, start by running the query in Listing 5-1, which will return a result for OrderId 21. As you can see, this order's status is showing as *returned*. Thankfully, we have two timestamps, OrderPlacedTimestamp and UpdatedAt, that give us a bit more insight into when the order was created and last updated. Although from these fields we know when the order was placed and returned, we are missing the details regarding when it was shipped and as such have lost valuable information.

Listing 5-1. Query to return the record for the returned order number 21

```
select
  OrderId,
  OrderStatus,
  OrderPlacedTimestamp,
  UpdatedAt
from raw_orders
where OrderId = 21;
```

If we had implemented snapshots to monitor for row changes in this table, we would have been able to retain a history of the status updates that this order had gone through and when those updates happened. This would unlock the ability to conduct a plethora of analysis on the data. An example would be trying to determine if the lag time between placing an order and when it was shipped is correlated to returns. Maybe we would be able to recommend that lowering the lag time would result in a decreased number of returns that we experience. But, without tracking historical changes, we can't answer these questions.

While you could use snapshots to track the changes to the *Orders* table, because of the velocity at which this data changes there are better ways to do this outside of dbt. We will discuss these potential issues with snapshots, and how you can overcome them, later in the chapter. For now, let's look at a second example for when you could use snapshots to monitor for row changes.

Example Two

In our previous example, we discussed using snapshots to track changes to the data in our *Orders* table, but the rate at which data changes in this table has the potential to be pretty high. While it isn't implicitly wrong to use snapshots to track changes on a table with high velocity data, there are better ways to do so, which we will discuss later. Let's review an example of a more ideal scenario using snapshots in practice to track row changes.

For this example, we will be working with the *raw_customers.csv* file. Recall from Chapter 2 that the *raw_customers* file in our project comes from a customer relationship management system (CRM). Knowing this, consider the following scenario. The salespeople in our business have access to make changes to customer records in the CRM and do so regularly when interacting with customers to ensure that we have up-to-date information about the customer. In the CRM, we keep track of the following information about a customer: name, phone number, email address, and shipping address. As you can imagine, for the majority of customers these attributes will change infrequently.

These infrequently changing, but still mutable, datasets are known as *slowly changing dimensions*. More specifically, dbt snapshots implement a Type-2 slowly changing dimension (SCD), meaning that history is tracked for each record in the source dataset. Note that the first word, slowly, implies that the velocity at which this data changes is infrequent. This is where snapshots are best utilized and are the most reliable, and customer-related data is a perfect example of this.

Note Slowly changing dimensions, made famous by Ralph Kimball, track changes in data from a source system. These changes happen slowly, but at somewhat uncertain intervals.

If you run the query in Listing 5-2, notice that this customer's record was created on January 2, 2023, but then updated on January 9, 2023. Given the current state of the data, we don't know what change was made, but let's say hypothetically that the update made was the customer's phone number being changed. The great thing with this is that this change happened seven days after the initial creation of this customer's record, which is certainly slowly changing enough for snapshots to capture this update.

Listing 5-2. Query to return the record for CustomerId 30

```
select
  CustomerId,
  Phone,
  CreatedAt,
  UpdatedAt
from raw_customers
where CustomerId = 30;
```

If we had implemented a daily snapshot on this table, we would have ended up tracking the changes to this customer's phone number. Table 5-4 gives you a visual representation of what the snapshot table would look like. Notice that there are now two records for *CustomerId* 30, where the first row captures the original phone number and the second row captures the updated phone number.

Snapshot Meta Fields

If you are new to working with Type-2 dimensions, the previous section should have provided you with the knowledge to understand them a bit better and understand when it makes sense to implement them via dbt snapshots. However, if you have previously worked with Type-2 dimensions, you are probably wondering how dbt implements the meta fields, such as valid date ranges, that you would expect in a Type-2 SCD. Fortunately, for us this is just another way that dbt helps streamline the development process because as part of the snapshot process a few typical metadata fields are added to the final table for us.

Why dbt Adds These Fields

When working with Type-2 SCDs, we know that we can have more than one row for each primary key in our source table. Referring back to Table 5-3 from earlier in the chapter, there are two records in this snapshot for *CustomerId* 100, but it would be much more simple for us to work with this data if we knew a few more things:

- Which of these two rows is the active one?
- What date range was each row active during?

While we could add some logic to our SQL to answer these questions, we don't need to because dbt adds four meta fields to every row in a snapshot:

- ***dbt_valid_from***: Timestamp from when the row was inserted. The way this field is calculated varies depending on the snapshot strategy used.
- ***dbt_valid_to***: Timestamp from when the row became inactive. This is effectively the same as the new active row's ***dbt_valid_from*** value.
- ***dbt_scd_id***: A hash value used by dbt to merge updates and inserts into the final table. You will likely never need to use this field.
- ***dbt_updated_at***: The timestamp from when dbt most recently updated the record. Also used internally by dbt's processes, and you shouldn't need to use this field.

So, dbt adds four meta fields to the table in your data warehouse, but as you can see, you will likely only use the first two. The ***dbt_valid_from*** and ***dbt_valid_to*** timestamps are where a lot of the value lies in Type-2 SCDs. These fields allow you to understand a point in time of the record you are working with. Table 5-4 expands on the example for CustomerId 100 from earlier in the chapter to show you what these meta fields look like.

Table 5-4. *Example of a snapshot with dbt meta fields:* ***dbt_valid_from*** *and* ***dbt_valid_to***

CustomerId	State	...	dbt_valid_from	dbt_valid_to
100	New Hampshire	...	2021-01-30 12:00:00	2023-01-31 05:30:00
100	Florida	...	2023-01-31 05:30:00	*null*

As you can see, we could now use this table in downstream models to conduct analysis based on a point in time, meaning that we know that the customer lived in New Hampshire from 2021 to 2023, but then moved to Florida. Additionally, you can query for the active record by filtering to only look at rows where the ***dbt_valid_to*** field is ***null***. This is just scratching the surface of how you can use Type-2 SCDs in practice, and we will discuss using these types of tables in downstream models later in the chapter.

Monitor for Row Changes

At this point, it should be clear that snapshots effectively build a Type-2 slowly changing dimension in your data warehouse. Of course, dbt makes this process fairly simple for you to do and doesn't require you to implement the typically needed boilerplate code for these types of transformations. Instead, you can just let dbt do this for you. Of course, this still requires that you understand how Type-2 SCDs work, and that is in part why we discussed a few examples of use cases for snapshots at the beginning of the chapter prior to diving into the implementation process.

dbt provides two different strategies for monitoring for row changes in your data sources: **timestamp** and **check** strategies. At a very high level, the timestamp strategy, as the name implies, monitors for record changes based on a timestamp in the source table. This is the most simple and accurate strategy, and because of this it should be your go-to strategy as long as you have a reliable timestamp in your source data. The second option, the check strategy, allows you to define a list of one or more columns that you want to check for changes. As you can imagine, this method will be a bit more complex to implement, but nonetheless we will cover how to implement both strategies and provide you with some best practices that you can use.

Timestamp Strategy

As we discuss the implementation process for both strategies, we will be using the *raw_customers* model, but we will start with the timestamp strategy. As mentioned before, this strategy is the easiest and most preferred way to configure snapshots. This strategy works by setting the ***updated_at*** configuration to the name of your timestamp column. This timestamp will then be used by dbt to monitor for row changes in your data source. Let's walk through how you can configure a snapshot using the timestamp strategy. For this example, we will be referencing this file: ***customers_snapshot_timestamp_strategy.sql***.

Setting up a .sql file for snapshots is slightly different from other types of materializations because you have to wrap your entire transformation in the snapshot Jinja block:

```
{% snapshot customers_snapshot_timestamp_strategy %}
...
{% endsnapshot %}
```

In your workflow with building snapshots, we recommend that you always start out by adding the snapshot block. Not only will this prevent you from forgetting to add the block and running into errors, but the block is also how you name your snapshots. Recall from Chapter 4 that with other materializations the name of the destination table is inherited from the name of your .sql file, unless otherwise configured explicitly. Snapshots are different in this sense because the name of the destination table is defined in the snapshot block. So, in our example earlier, this table will be named ***customers_snapshot_timestamp_strategy***.

Next, you should add in the select statement for this model. Remember that you should almost always snapshot a single source that you have configured in your project's ***sources.yml***. However, for the purpose of our example, we will be selecting from a single ***ref*** because our sources are made up of seed files. Additionally, we don't recommend that you add any transformations into your snapshot query, but instead just ***select*** * from the source table. This keeps your snapshots simple, and you can always add transformations in downstream models. At this point, your snapshot file should look like this:

```
{% snapshot customers_snapshot_timestamp_strategy %}
select
  *
from {{ ref('raw_customers') }}
{% endsnapshot %}
```

The two steps that we just walked through should be done for every snapshot that you build, regardless of whether you are going to use the timestamp or check strategy. The third, and final, step is where we will configure the specifics about our snapshot so that dbt knows how we want to monitor for row changes. There are four configurations that you have to define when you are using the timestamp strategy for a snapshot:

- ***target_schema***: The schema in your database where dbt should materialize the snapshot
- ***strategy***: The strategy to use for the snapshot (timestamp or check)
- ***unique_key***: The field, or expression, that should be used by dbt to uniquely identify a record in the source table
- ***updated_at***: The timestamp that dbt should use to identify new and updated rows

To finalize our first snapshot example, we will need to identify what values we should set each of these configurations to. We know that the primary key of our source table is ***CustomerId***, so we will set that to be the ***unique_key***. Additionally, we have an ***UpdatedAt*** field in our source table, so we set that to ***updated_at***. It just so happens that our timestamp field is similar to the dbt config's name, but your timestamp field could, and likely will, vary from this in practice. Of course, the ***strategy*** config will be set to ***timestamp***, but we will address the ***check*** strategy in the next example. Lastly, we want to materialize our snapshots in an isolated schema, so we will set the ***target_schema*** config to render these tables in a schema called ***snapshots***. We can now bring this all together and add a ***config*** block to the top of our .sql file. With the configurations set like we did as follows, our first snapshot example is completed:

```
{% snapshot customers_snapshot_timestamp_strategy %}
{{
   config(
     target_schema='snapshots',
     unique_key='CustomerId',
     strategy='timestamp',
     updated_at='UpdatedAt'
   )
}}

select
  *
from {{ ref('raw_customers') }}
{% endsnapshot %}
```

Now that we have walked through the process for building this snapshot, for the last step you will need to execute the ***dbt snapshot*** command to have dbt to build this as a table in your database. Let's make our command a bit more constrained and use the --***select*** operator so that we only run this snapshot. We will get into more details regarding what dbt is actually doing behind the scenes later in the chapter, but for now just run the following command to execute the snapshot:

```
dbt snapshot --select customers_snapshot_timestamp_strategy
```

You can validate that this worked by checking that you received a success message from dbt as well as checking your warehouse for the presence of this table:

```
dbt_learning.snapshots.customers_snapshot_timestamp_strategy
```

Meta Fields with the Timestamp Strategy

Now that you have run this snapshot and the table has materialized in your data warehouse, let's take a look at the meta fields that dbt added to this table. Recall from earlier in the chapter that dbt will add fields to give you record validity time ranges using the ***dbt_valid_from*** and ***dbt_valid_to*** meta fields. We want to call out how dbt sets the ***dbt_valid_from*** field because it differs between the two strategies that you can use for snapshots.

When you build a snapshot using the timestamp strategy, dbt will use the value from the timestamp field that you set for the ***updated_at*** config. In our example, this has been configured to use the ***UpdatedAt*** field from the ***ref_customers*** model. Let's walk through an example where we look at an active row in the snapshot table, simulate an update to the source, rerun the snapshot, and see how the meta fields are affected.

First, you should already have this table built in your warehouse, so if you run the query in Listing 5-3, it will return the result seen in Table 5-5.

Listing 5-3. Query to return the records for CustomerId 1

```
select
  CustomerId,
  Phone,
  UpdatedAt
  dbt_valid_from,
  dbt_valid_to
from dbt_learning.snapshots.customers_snapshot_timestamp_strategy
where customerid = 1
```

Table 5-5. *Snapshot meta fields when using the timestamp strategy*

CustomerId	Phone	UpdatedAt	dbt_valid_from	dbt_valid_to
1	1-394-354-2458	2022-12-21 14:08:36.464	2022-12-21 14:08:36.464	*null*

Notice that in Table 5-5, the ***dbt_valid_from*** field is using the same value as ***UpdatedAt*** because, as we called out earlier, this meta field is set based on how you defined the ***updated_at*** config. This is easy enough to understand, but let's examine how dbt updates the meta fields when the source data is changed. Since we aren't working with live data in these examples, you can instead run this query to simulate a change to the ***raw_customers*** table:

```
update dbt_learning.public.raw_customers
set Phone = '1-394-354-2456',
    UpdatedAt = '2023-01-01 12:06:34.465'
where CustomerId = 1
```

This query is going to update the phone number and the update timestamp for customer 1. While we would never recommend updating your production data like this, it is convenient for us to do in our learning environment. With that said, now that we have updated the phone number for customer 1, go ahead and run the snapshot command again for this model:

```
dbt snapshot --select customers_snapshot_timestamp_strategy
```

Once that completes successfully, you should be able to rerun the query in Listing 5-3, but notice that there are now two rows in the table for customer 1. One active record and one inactive record are now present in the table, and your result should look like the rows in Table 5-6.

Table 5-6. *Snapshot meta fields after this customer's phone number was updated*

CustomerId	Phone	UpdatedAt	dbt_valid_from	dbt_valid_to
1	1-394-354-2456	2023-01-01 12:06:34.465	2023-01-01 12:06:34.465	*null*
1	1-394-354-2458	2022-12-21 14:08:36.464	2022-12-21 14:08:36.464	2023-01-01 12:06:34.465

As you can see, dbt was able to pick up the updated phone number for this customer, as well as update the meta fields. As you can see in Table 5-5, the record that was previously active now has a value in the ***dbt_valid_to*** column. Notice that this value is the same as the active row's ***dbt_valid_from*** value. The key takeaway here is that when using snapshots dbt will update the meta fields using whatever column you set in the ***updated_at*** config.

Note In the next section, we will discuss the check strategy, but the process for dbt setting the meta fields is almost the same, so we won't cover this information again. The key difference is that when using the check strategy the ***dbt_valid_from*** field is set using the system's current timestamp at run time.

Check Strategy

While the timestamp strategy should be your preferred method for implementing snapshots, not every data source is going to have a reliable timestamp that you can use. When you need to snapshot data sources like this, you can use the check strategy. With this strategy, you configure a list of one or more columns that dbt should check for changes when your snapshot runs.

Caution The check strategy generates a query that compares every column that you configure to check for changes using the ***or*** keyword in the ***where*** clause. If you are checking for changes in too many columns, you may end up with long-running snapshots.

Unfortunately, the check strategy is slightly less straightforward to implement compared to the timestamp strategy because you have to make considerations about which column(s) you want to monitor to trigger a new record in your snapshot table. In our experience, there are three approaches that you can take to implement snapshots using the check strategy. While using the check strategy isn't very ideal to begin with, we do think that these three approaches have a clear progression from least ideal to most ideal, and we will cover them in this order.

- Check all columns
- Check columns in a list
- Check a hash column

Before we dive into each approach, take a look at the following configurations that you need to set when using the check strategy. Notice that there are subtle differences between these configs and the ones that you have to set when using the timestamp strategy:

- ***target_schema***: The schema in your database where dbt should materialize the snapshot
- ***strategy***: The strategy to use for the snapshot (timestamp or check)
- ***unique_key***: The field, or expression, that should be used by dbt to uniquely identify a record in the source table
- ***check_cols***: A list of columns that dbt should check to identify row changes. Can alternatively be set to "***all***" to check every column in the source

Check All Columns

If you're in a situation where you need to use the check strategy, and don't know where to start, the simplest approach is to configure the snapshot to check for changes across every column in your source. The steps for setting up the check strategy using this approach don't differ much from the timestamp strategy example that we covered earlier in the chapter. The key differences come from the config block at the top of our file. For this example, you can reference this file: ***customers_snapshot_check_strategy__all.sql***. If you want to build this snapshot in your data warehouse, you can run this command to do so:

```
dbt snapshot --select customers_snapshot_check_strategy__all
```

Notice that in the config block, compared to the config block in our timestamp strategy example, there are only a couple of differences:

```
{{
  config(
    target_schema='snapshots',
    unique_key='CustomerId',
    strategy='check',
    check_cols='all'
  )
}}
```

As with all of our snapshot examples, we are materializing these tables in the ***snapshots*** schema in our data warehouse. Since we are still dealing with the ***raw_customers*** models as our source for this snapshot, the unique key hasn't changed in the configuration either. The subtle, yet important, differences are in the last two configurations where we set the strategy to ***check*** and we set the ***check_cols*** config to ***all***.

Note ***all*** is a reserved string within the ***check_cols*** config that lets dbt know that you want to check for row changes across all columns.

While this approach is very simple, the query that dbt generates to capture these changes becomes increasingly complex as the number of columns in your source grows. This is not a problem from an implementation standpoint; as a matter of fact, this is a great example to demonstrate the power of dbt. All that you did was simply define the ***check_col*** config, and dbt generates a fairly complex query to capture these changes. It is really awesome that dbt can do this, and it is great while things are working. But, you will run into issues with this approach when you inevitably have to debug your snapshot and look at the query that dbt compiles for you.

With an understanding of how to configure the check strategy using this approach and the potential pitfalls, you are probably wondering when you should use the check all columns method. In our experience, it is best to use this approach if your use case meets these criteria:

- There is not a reliable timestamp in the source table to monitor for updates. If there is, use the timestamp strategy instead.
- The source table has a small number of columns.
- You don't expect that the number of columns will increase very much over the long term.
- You need to monitor for changes across all of the columns in the source table.

If you've checked off all of these criteria, you will likely get away with using the check all method. This approach allows you to keep things simple, and you can focus more on developing and less on overengineering a Type-2 dimension. With that said, there are two more approaches that you can take to implement the check strategy in a snapshot.

Check Columns in a List

In our prior example, we set up a snapshot using the check strategy using the ***all*** keyword to tell dbt to check every column in our source for row changes. However, we may not always need to check for changes to every column. There may be times where you know that a field in your dimension will never change, or perhaps you simply don't care if a field changes. These are both situations where it makes sense to explicitly list the columns to check for changes in.

For this example, you can refer to this file to see how we configured the snapshots: ***customers_snapshot_check_strategy__list.sql***. Additionally, to build this snapshot in your warehouse, run this command:

```
dbt snapshot --select customers_snapshot_check_strategy__list
```

Compared to the check all columns method that we just covered, the configurations in this snapshot differ slightly. Everything in the config block will be the same with the exception of the ***check_cols*** config. Instead of setting this to the ***all*** keyword, we will pass in a list of strings that correspond to column names in the source. Let's say that we only want dbt to monitor for row changes based on the following columns:

- Name
- Phone

If these are the only columns that we care about monitoring changes to, then all that we need to do is update the config block to look like this:

```
{{
   config(
     target_schema='snapshots',
     unique_key='CustomerId',
     strategy='check',
     check_cols=['Name', 'Phone']
   )
}}
```

Now when we run the snapshot command, dbt will only look for differences in these two columns. If at some point in the future we decide that we need to include the ***email*** column in our config, you can simply add it to the list of columns to check like this:

```
{{
   config(
     target_schema='snapshots',
     unique_key='CustomerId',
     strategy='check',
     check_cols=['Name', 'Phone', 'Email']
   )
}}
```

As you can imagine, it is useful to have control over which columns are being used to trigger new rows in your snapshots. However, this does come with some added responsibility and maintenance from the developer's point of view. This alone presents a few issues compared to the check all columns method.

Consider a scenario where a new column has been added to the source table; you as the developer will need to make the decision on whether or not you want to have this new field to be included in the configured list. But, before you can even make this decision, you must be aware that the new column was added to the source. While data engineers and software engineers in many organizations are striving to work more closely together via defined schemas and data contracts, not every organization has reached this level of maturity. So, you will need to consider how your organization works in order to tackle this.

Not to say that this approach isn't a good one, we simply want you to be aware of the considerations you need to make when implementing this solution. As a matter of fact, in our opinion, this method is still a better approach than checking all of the columns because it removes some of the complexity from the query that dbt generates to detect row changes. Be intentional with the method you select, and don't just default to checking all columns because it is easier.

Check a Hash Column

Let's wrap up this section by walking through an example of how we can use a hash column to detect row changes in our source. There are a few reasons why this is our favorite approach to implementing snapshots if we have to use the check strategy:

1. You can take advantage of one of dbt's most powerful features, Jinja.
2. You can monitor for changes in every column via one string value.
3. The query that dbt generates to monitor for row changes is simplified because it is only comparing a single column.

Note For this example, we will be referencing this file: ***customers_snapshot_check_strategy__hash.sql***.

Before we cover how to set up a snapshot using this method, let's first discuss what we mean by the term "hash column." In the context of our example, a hash column refers to a single string column that is the result of concatenating all of the columns in the record and applying a hashing algorithm to that concatenated string. Typical hashing algorithms that you will run across in data warehousing are MD5 and SHA-2.

Caution The MD5 algorithm produces a shorter string, which increases the chance of hash collision, compared to the longer SHA-2. This is only an issue if you expect your table to have an extremely large record count.

Take a look at Listing 5-4 for an example of how we can generate a hash column using the MD5 algorithm.

Listing 5-4. Example of how to generate a hash value of all columns in the source

```
select
  *,
  md5(cast(coalesce(cast(CustomerId as text), null) || '-' ||
  coalesce(cast(Name as Text), null) || '-' ||
  ...
  coalesce(cast(UpdatedAt as Text), null) as text)) as HashDiff
from {{ ref('raw_customers') }}
```

We generate the hash by casting all of our columns in the source to the same data type and then concatenating them into a single string. Lastly, the MD5 algorithm is applied to this string, and we end up with a 32-character string value that is unique to each row. You can see an example of the result in Table 5-7. Keep in mind that we are using Snowflake, so you may need to tweak this slightly to fit your SQL dialect. Fortunately, dbt has a solution for this already that we will discuss as we go through the example.

Table 5-7. *Example of a hash column, named HashDiff*

CustomerId	Name	...	HashDiff
1	Genevieve Trevino	...	74c6f4ac156965f202c7c732d3c577ab

To start implementing a snapshot using the check strategy in combination with the hash column method, we need to add a transformation to our snapshot file to generate the ***HashDiff***. We can do this in two ways, the first being generating the hash ourselves like we did in Listing 5-4, or we can utilize dbt macros to generate this column for us. We are going to cover macros in great detail in Chapter 6, but now is a great time to start to uncover the power that macros and Jinja bring in terms of simplifying your code.

If you refer to the file for this example, ***customers_snapshot_check_strategy__hash.sql***, or the following code snippet, you will see we are using Jinja functions and macros to generate the ***HashDiff*** column in our snapshot:

```
select
  *,
  {{ dbt_utils.generate_surrogate_key(adapter.get_columns_in_
  relation(ref('raw_customers'))|map(attribute='name')|list) }} as HashDiff
from {{ ref('raw_customers') }}
```

Since this chapter isn't meant to cover Jinja and macros at a granular level, let's instead gain a high-level understanding of what is happening in this statement to generate the hash column. If we look at the innermost part of this statement, in Listing 5-5, the first thing that is happening is we are using the ***get_columns_in_relation*** method from the ***adapter*** Jinja wrapper. This method returns an object with metadata about the columns in a relation. The method takes one input parameter, the relation, which can be passed in via a dbt function to reference a source or upstream model. While you could pass in a static table name, as with any other time you are writing code using dbt, you should try to avoid doing this. Since this method provides more information than just the column name, we add a bit of Jinja to ***map*** to just the column names and then convert those to a list.

Listing 5-5. Get a list of column names from the raw_customers model

```
adapter.get_columns_in_relation(ref('raw_customers'))|map(attribute=
'name')|list
```

We do this because the `dbt_utils.generate_surrogate_key` macro expects a list of strings that represent column names. The ***generate_surrogate_key*** macro from the ***dbt_utils*** package then concatenates all of these columns and applies the MD5 algorithm to them to create a hash column for us. This is effectively doing the same thing that we did manually in Listing 5-4, but instead we use the power of Jinja and macros to take some of this heavy lifting off our shoulders and make our code more DRY. With that said, we will cover packages, such as ***dbt_utils***, in much greater detail in Chapter 6.

Now we have a column in our transformation that creates a hash of all of the columns in the table that is the source for the snapshot. In our case, that is the ***raw_customers*** model. We can now update the config block such that the ***check_cols*** configuration only looks at the ***HashDiff*** column to monitor for row changes:

```
{{
   config(
     target_schema='snapshots',
     unique_key='CustomerId',
     strategy='check',
     check_cols=['HashDiff']
   )
}}
```

So, why go to all of this effort when dbt has the ***all*** keyword built in for use with the check strategy? There are a few reasons that this is a better solution than check all:

- The query dbt generates will be more simple as it is only checking for changes in one column, but you still get the benefit of tracking changes to some/all columns.

- You *could* see improved performance with this method over the check all columns method. We use the term *could* loosely here because there are so many factors that play into performance.

We recommend that if you need to implement a snapshot using the check method, you always use this approach. Not only do you get the benefits mentioned earlier, but it also standardizes how you implement snapshots using the check method. You could even template this so that you can spin up a new snapshot in just a few minutes by swapping out the source references.

Additional Configurations

While snapshots have several required configurations, you can also add other configurations to alter the behavior of your dbt snapshots. In this section, we will discuss how you can configure the snapshot behavior for handling hard deletes from the source, as well as building your snapshot in a specified database.

Although we are covering these snapshot-specific configs, there are several other configs that will be familiar to you from models, seeds, etc. If you need to do so, you can set any of these other optional configurations:

- ***enabled***
- ***tags***
- ***pre-hook***
- ***post-hook***
- ***persist_docs***
- ***grants***

Hard Deletes

In our previous examples, we didn't address the issue of hard deletes that is often present when working with Type-2 SCDs. There is an optional configuration that you can set for snapshots that will monitor your source for deleted records and set them to inactive in your dimension table.

Note dbt will only monitor for hard deletes. If your source table uses a soft delete method, you will need to handle those in a downstream model.

If the source that you are snapshotting has instances of hard deletes, you should consider enabling the ***invalidate_hard_deletes*** config by setting it to ***True*** as follows:

```
invalidate_hard_deletes=True
```

dbt adds a few extra steps to the snapshot process to monitor for and deactivate deleted records. The first step that dbt takes is to identify which rows have been deleted from your source table. This is achieved by using the common left anti-join SQL pattern. What this means is that dbt will left join the source table to the snapshot table and check for nulls in the source table. The join is built using your configured ***unique_key***. Figure 5-1 provides you a visual representation of how these hard deletes are identified by dbt snapshots. Effectively, dbt is querying for records that are in the snapshot, but no longer in the source table.

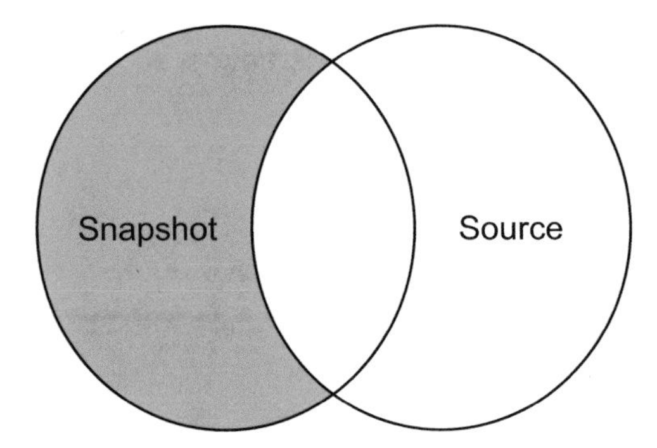

Figure 5-1. *Visual representation of a left anti-join*

Once dbt has identified the set of rows that have been deleted from the source, it will update these in the snapshot table by setting the ***dbt_valid_to*** meta field to the timestamp from when the snapshot was run. Contrary to inserts and updates for snapshots, this field has to be set using the current timestamp and not one from the source because the record no longer exists. This could be solved using an alternative approach to tracking history in tables, and we will discuss that later in the chapter when we cover the potential issues with snapshots.

Target Database

Although your project has a default database configured in your profile, there are times when you may need to materialize tables into a different database. This is very useful when you start using environment variables to be able to run your project in different environments. While other dbt assets such as models and seeds use the ***database*** config, to achieve this same behavior with snapshots you will need to use the ***target_database*** config.

Setting Snapshot Configurations

Throughout the examples in this chapter, we have configured snapshots using the config block at the top of our .sql files, but similar to models and seeds, you can also configure snapshots in ***.yml*** files. This can be done by either defining the configs in the ***dbt_project.yml*** or within a config resource property within some other ***.yml*** file.

Tip To keep your project organized, we set up ***<directory>_schema.yml*** files. For snapshots, this translates to ***snapshots_schema.yml***.

The hierarchy in which configs are applied to snapshots follows the pattern that you can see in Figure 5-2. What this is showing is that dbt will use the configs at the lowest level of configuration. For example, if you define a ***target_database*** in the ***dbt_project.yml*** and within your snapshot's ***.sql*** file, then the configuration set in the ***.sql*** file will take precedence.

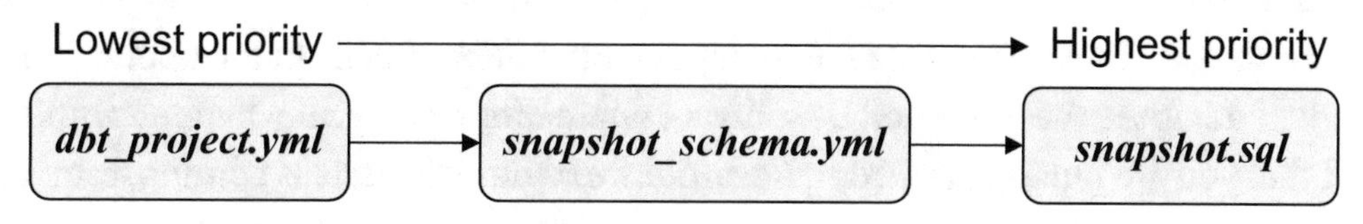

Figure 5-2. *Configuration hierarchy for snapshots*

Setting Required Configurations

While dbt will navigate the hierarchy we discussed to determine which configurations to use for a snapshot, we want to make sure you're aware of a nuance of where you should set the required snapshot configurations. Recall from earlier in the chapter the required configs include

- ***target_schema***
- ***strategy***
- ***unique_key***
- ***check_cols***
- ***updated_at***

The nuance is that you should only set these configs in either the config block at the top of the snapshot's ***.sql*** file or directly within the ***dbt_project.yml*** file. You should avoid setting these in the config resource property of any other ***.yml*** files because they may not work when set this way. This happens because when dbt parses your project, it looks for the required configs to be set at the project level or directly in the snapshot file, so if either of these aren't set, dbt will throw an error. You could work around this by always setting a default value in the ***dbt_project.yml***, but that is less than ideal and will likely cause you some headaches when you need to debug.

How dbt Handles Schema Changes

Oftentimes, when data professionals are working with raw source data, we experience schema changes. These changes tend to happen upstream directly in the application or product database that is owned by software engineers, or they may happen within an API that you are pulling data from. Regardless of where the data is coming from, you should understand how your data pipeline handles schema changes, so that when things inevitably change you aren't left scrambling to fix your broken pipeline.

What do we mean by schema changes anyway? We like to think of schema changes as any alterations to the structure or format of our source data, such as

- Changing table names
- Adding or dropping tables

- Changing column names
- Adding or removing columns
- Data type changes
- Keys changing

All of these schema changes are important to understand, and anticipate, in your data pipelines, but within the context of snapshots, we will cover just a few of these. We will focus on column-level changes and how dbt handles new columns, removing columns, and data type changes within existing columns.

With any of the changes that we are going to discuss, dbt will only pick these changes up if you are selecting the changed or new column in your query. For example, if your snapshot query looks something like `select * from {{ some_source }}`, then anytime a column is added, altered, or removed dbt will be aware of it. You could list the columns that you want to select in your query, but you may run into failures if a column you are selecting is removed from the source, or if a new column is added to the source, dbt simply won't know to add it to the snapshot. There are trade-offs when selecting one method over the other. For example, when selecting all columns

- dbt will handle schema changes better.
- But, you could incur performance issues if your source has many columns, and you don't need to snapshot all of them.

Consider these trade-offs when determining if it is best to select all or specific columns in your snapshot query. Now, let's dive into how dbt handles schema changes with respect to snapshots.

Adding New Columns

It's likely that one of the most common schema changes you will encounter is new columns being added to your data source. This typically happens when there is a need for data upstream wherever you are pulling this data from, but it may have also been added at your request specifically for an analytics use case. Either way, you will want to know how dbt handles new columns being added to your snapshot query.

Luckily, when dbt detects a new column in your snapshot query, it will add this new column to the destination snapshot table in your data warehouse. Take a look at Table 5-8 to see how this would impact the destination table of our snapshot if a new column ***DateOfBirth*** was added to the raw customers source file and the statement `select * from {{ some_source }}` was used.

Table 5-8. *Example of a new column being added to a snapshot destination table*

CustomerId	Name	...	UpdatedAt
3	Leslie Rollins	...	2023-02-24 20:07:47.418

CustomerId	Name	...	UpdatedAt	DateOfBirth
3	Leslie Rollins	...	2023-02-24 20:07:47.418	1989-06-04

As you can see, dbt would pick this field up and append it to the end of our snapshot destination with no needed intervention on our part since our query is a select all. As mentioned before, if we had explicitly listed the columns to select, the new ***DateOfBirth*** field would not have been picked up without us adding the column to the select statement in the snapshot file.

Removing Columns

Another example of a schema change is when a column is removed from your snapshot query. This could be a result of you removing the column from your query or the column being removed from the data source itself. Regardless of why the column is no longer a part of the snapshot query, you will want to understand how dbt handles this kind of schema change.

When a column is removed from a snapshot query, dbt will handle this change by no longer adding values to the column in the destination table. Fortunately, dbt will not delete the column, so you will still retain the history that you have tracked for that column.

Note A column will not be deleted from a snapshot destination table even if that column is removed from the source.

Table 5-9 gives an example of what this would look like if we removed the ***Phone*** field from our snapshot query. In this example, customer 501 is a newly added record, and as you can see the ***Phone*** field still exists, but it is ***null*** for this new customer record if the statement `select * from {{ some_source }}` was used.

Table 5-9. *Example of a snapshot destination table behavior when a column is removed from the source*

CustomerId	**Name**	**Phone**
3	Leslie Rollins	1-584-147-6573
...	...	...
501	Lionel Messi	*null*

Data Type Changes

The final type of schema change that you should be mindful of is what dbt will do when data types change in the source that you are snapshotting. This is a bit different than the other changes that we have discussed because dbt will handle data type changes differently depending on what the data type is in the snapshot destination table.

Typically, dbt will not change the data type in your destination even if the data type changes in the source. For example, if you have a column with a data type of *date*, but upstream the data type is changed to a *timestamp*, dbt will not attempt to reconcile this change in the snapshot destination table. Therefore, all of the values that are now timestamps will be converted to dates as they are inserted into your snapshot table. You should keep this in mind if you know that a data type in the snapshot's source is changing. It could also be useful for you to set up some monitoring and alerting so that you know when these types of changes happen in your raw/source data.

There is one particular scenario where dbt will attempt to update the data type of a column in the destination table. If your destination column is of some string type and the size of that string column increases upstream, dbt will increase the size of the string column in the snapshot table. For example, if you have a column, *BusinessName*, of type *varchar(255)* and upstream it is converted to a *varchar(500)*, then dbt will attempt to follow suit and expand the column size.

Using Snapshots in Downstream Models

Up to this point, we have dedicated this entire chapter to showing you the different methods for implementing snapshots, but let's now shift gears and take a look at how you can actually use snapshots in development of downstream models. Let's take a look at an example of how snapshots can be queried for use in downstream models. For this example, and throughout this section, we will be using the ***fct_orders.sql*** file.

Before we start the example, recall from earlier in the chapter that you should almost always treat snapshots the same as you do sources. By this, we mean that you should have no, or very limited, transformations in your snapshot. If there are some light transformations that you need to do to the data, we recommend that you do that in a staging model. Here, you can rename columns, cast data types, or do some other light transformations. However, to keep our example simple, we simply will be using a reference to the snapshot itself in a downstream model. Just keep in mind that in practice you may need a staging model in between these steps.

Referencing Snapshots

Snapshots are referenced downstream using the `{{ ref() }}` function. This should be familiar to you at this point because this is the same way that you reference models and seeds. In our example, we have started building a fact table that is comprised of two assets:

1. ***models/staging/furniture_mart/stg_furniture_mart_orders.sql***
2. ***snapshots/customers_snapshot_timestamp_strategy.sql***

This is a simple transformation that joins these two tables together to incorporate order-related information and the customer's name to the model. As you can see in the following example, both of these dbt assets are referenced using the `{{ ref() }}` function even though one is a model and the other is a snapshot:

```
select
 ord.OrderId,
 ord.CustomerId,
 cus.Name,
 ord.SalesPerson,
 ord.OrderStatus,
```

```
 ord.OrderPlacedTimestamp
from {{ ref('stg_furniture_mart_orders') }} as ord
left join {{ ref('customers_snapshot_timestamp_strategy') }} as cus
 on ord.CustomerId = cus.CustomerId
```

If you are interested to see what the results of this query yield so far, you can run this command to have dbt build the model:

```
dbt run --select fct_orders
```

However, once this has been built successfully in your data warehouse, you should next query the results of this table to check the grain of the table to make sure that the transformation is behaving as expected. You can do this by checking that the number of orders in the destination table is equal to the distinct number of orders because we expect to have one row per order in this table. Do this by running the following query:

```
select
  count(OrderId) as NumberOfRecords,
  count(distinct OrderId) as DistinctNumberOfOrders
from dbt_learning.public_operations.fct_orders
```

After checking this query, you will notice that there are 1005 records in this table, but only 1000 distinct orders, so we have an issue with our transformation somewhere. This join fanout is being caused by the snapshot table. Remember that snapshot tables track history and as such can have more than one record per the unique id you are tracking. In this case, the snapshot table is tracking changes to Customers, so the table can have more than one record per customer. Recall from earlier in the chapter, we made updates to ***CustomerIds*** 1 and 30. Fortunately, when working with Type-2 dimensions, there are a few ways that we can solve this issue when using snapshots downstream models:

1. Query for only the active snapshot records.
2. Query for a snapshot record at a point in time.

Query for the Active Record

The most straightforward way that you can reference snapshots without having to worry about join fanout is to query for only the active record. Recall the meta fields that dbt will always add to snapshot tables the first time the snapshot is run. One of those meta fields in particular, ***dbt_valid_to***, will assist us in retrieving only the active row.

Let's update the example from the previous section to add a where clause to limit the returned records to be only those that have a ***null*** value in the ***dbt_valid_to*** column because this tells us that this is the active record. The following is what the query should look like in the ***fct_orders*** model after this change:

```
select
  ord.OrderId,
  ord.CustomerId,
  cus.Name,
  ord.SalesPerson,
  ord.OrderStatus,
  ord.OrderPlacedTimestamp
from {{ ref('stg_furniture_mart_orders') }} as ord
left join {{ ref('customers_snapshot_timestamp_strategy') }} as cus
  on ord.CustomerId = cus.CustomerId
where cus.dbt_valid_to is null
```

Then, run and validate the results the same way that we did before. Notice that from the validation query, the number of records in the table does equal the distinct number of orders, 1000. This solves the join fanout issue, but this method doesn't provide data about the customer based on when the order was placed. It instead tells you about the current state of the customer record in the snapshot table. Depending on your use case, this may be fine, or even ideal, but let's look at how we would want to structure this transformation if we wanted to query the customer snapshot table at a point in time.

Caution Be careful using the active row method if you have to deal with hard deletes in your sources and have configured your snapshot to `invalidate_hard_deletes=True` because there is not an active record to be returned. As such, orders associated with hard-deleted customers would not be returned.

Query for a Record at a Point in Time

If you need to return a snapshot record from a point in time within a dbt transformation, you can do that with a time range join using the snapshot meta fields. This method is our preferred method, when performant, because it keeps your models more idempotent than using the active record method. This is because every time you run the downstream model using the point-in-time method, the same result will return for any given record. However, the active record method will not always return the same result. In fact, it will change every time the corresponding record is updated in the snapshot.

To use the point-in-time method, you will need to remove the where clause that we added in the prior example and instead update the join criteria. You should update the join criteria to also join on a timestamp that is between ***dbt_valid_from*** and ***dbt_valid_to***. The timestamp you use in this criteria should make sense in the context of our transformation, but in this example we use the ***OrderPlacedTimestamp*** to return the customer record at the time the order was placed:

```
select
 ord.OrderId,
 ord.CustomerId,
 cus.Name,
 ord.SalesPerson,
 ord.OrderStatus,
 ord.OrderPlacedTimestamp
from {{ ref('stg_furniture_mart_orders') }} as ord
left join {{ ref('customers_snapshot_timestamp_strategy') }} as cus
 on ord.CustomerId = cus.CustomerId
 and ord.OrderPlacedTimestamp between cus.dbt_valid_from and
 ifnull(cus.dbt_valid_to, '2099-12-31'::timestamp_ntz)
```

Before you validate these results, also notice in the join criteria that we checked to see if the ***dbt_valid_to*** column is null and plugged in a default value of some date arbitrarily in the future. We have to do this in case the active record is in fact the correct point-in-time record as well. Lastly, as you did with the other two examples, you can run the same validation query and see that this method also solves that fanout issue.

Potential Issues with Snapshots

Recall Example One from earlier in this chapter under the "When to Use Snapshots" section. In this example, we discussed how you could implement a dbt snapshot to track row changes in your data warehouse for a data source that comes from a relational database such as PostgreSQL or SQL Server.

As you can see, implementing a dbt snapshot would have served us well by retaining the history of these changes. However, there are two caveats to using snapshots that we want you to be aware of when determining your approach for tracking history of a data source such as our *orders* table. First, when snapshotting a data source with dbt, you will need to understand the velocity at which data is generated, updated, and deleted because these factors will directly impact the second caveat: snapshot frequency. You need to consider both of these factors when using snapshots because if your data source has high velocity, but you don't take snapshots at high enough frequency, you run the risk of potentially missing data.

Take Table 5-10 as a hypothetical example of the full history of order 21 (from Listing 5-1) that we can use to analyze the velocity at which the status is updated and how snapshot frequency will impact the state of our history tracking in the data warehouse. This example is fairly straightforward; the order is first placed on March 6, 2020, and goes through four updates where the *OrderStatus* changes a few times. An interesting situation has happened here where the *OrderStatus* toggles between shipped and placed a few times.

Table 5-10. *Example of the full history of OrderId 21*

OrderId	OrderStatus	OrderPlacedTimestamp	UpdatedAt
21	placed	2020-03-06 08:26:31.000	2020-03-06 08:26:31.000
21	shipped	2020-03-06 08:26:31.000	2020-03-07 22:51:34.000
21	placed	2020-03-06 08:26:31.000	2020-03-07 22:52:00.000
21	shipped	2020-03-06 08:26:31.000	2020-03-10 07:54:01.000
21	returned	2020-03-06 08:26:31.000	2020-03-19 19:27:41.004

In this example, let's assume that an operational error occurred where an employee incorrectly marked this order as *shipped* on March 7, 2020, and then almost immediately back to *placed*. Although this was something that happened in error, it would be advantageous for us to track this in the data warehouse still.

Now that we have an understanding of how this record has been updated over time, let's consider the frequency at which we run the ***dbt snapshot*** command against this table. Oftentimes, snapshots are being run against source tables once per day or even once per hour. We will need to determine the correct frequency, or take an alternative approach, to prevent data loss.

Imagine that we are running the ***dbt snapshot*** command one per hour at the top of the hour. Even with this frequency, we will end up not tracking the full history of this record, and instead we will end up with records that look like the results in Table 5-11. Notice that when comparing this to Table 5-10, we are missing the record where the *OrderStatus* was updated to *shipped* in error, but we did capture the correction to that error. Why does this happen?

Table 5-11. *OrderId 21 history from snapshot with missing data*

OrderId	OrderStatus	OrderPlacedTimestamp	UpdatedAt
21	placed	2020-03-06 08:26:31.000	2020-03-06 08:26:31.000
21	placed	2020-03-06 08:26:31.000	2020-03-07 22:52:00.000
21	shipped	2020-03-06 08:26:31.000	2020-03-10 07:54:01.000
21	returned	2020-03-06 08:26:31.000	2020-03-19 19:27:41.004

The reason this happens is because since we are taking our snapshot at the top of the hour, we are only going to capture the current state of the record from the source. Notice that in the 22nd hour on March 7, 2020, order 21 experienced two updates. Snapshots in dbt have no way to track intermittent changes, and as such we don't have any history of the accidental update where an employee accidentally marked this order as shipped.

There are a few ways to solve this issue with this simplest solution being to simply increase the frequency at which you are snapshotting the source table. However, you would need to snapshot the source table nearly every minute, or more, for changes like this to be captured. As such, if you need to track changes to your source data that has a high velocity, it may be advantageous for you to explore other options such as transaction log-based change data capture or event streams.

Summary

In this chapter, we introduced you to dbt snapshots and when it makes sense to use them in practice. We walked through two examples of what it may look like to use snapshots given sources with different levels of volume and velocity. This provides you the basic understanding of how snapshots are useful, but also when it is best to use them.

We then moved into how you can identify when snapshot records were updated or how to identify the active record using the meta fields added by dbt. These are fields that you commonly see in Type-2 slowly changing dimensions, but fortunately dbt adds these automatically to snapshots. Keep in mind that there are four different meta fields that dbt adds in, but in practice you will mostly work with the ***dbt_valid_from*** and ***dbt_valid_to*** fields.

From here, we moved into discussing the different strategies that you can use to implement snapshots. We first covered the timestamp strategy which uses a trustworthy timestamp from a source table to capture updates within the snapshot destination table. This strategy is the most simple and effective way to implement snapshots and as such should always be your default unless you don't have a timestamp to base updates off of. In instances where you don't have a timestamp to use in snapshots, you can instead use the check columns strategy. There are several different approaches you can take to implement this strategy. The method that you choose to implement should be influenced by the volume and velocity of your data, as well as the number of columns in your source table that you are snapshotting.

Once you have a snapshot implemented, you have to configure it so that dbt knows how to build it in your warehouse. This is done through a set of required configurations, but there are optional configurations that you can set as well. The optional configuration with the greatest implications is `invalidate_hard_deletes`.

Snapshotting data from upstream sources tends to have some level of uncertainty associated with it because analytics engineers and data engineers typically do not own these upstream sources. As such, the schemas to these sources often change with little or no communication to the analytics teams. For this reason, snapshots need to be resilient to upstream schema changes, and so we covered how dbt handles schema changes in snapshots.

Up to this point, we covered only how to implement snapshots, but all of that effort would be wasted if we didn't also discuss how to use them downstream. So, we covered an example of incorporating a snapshot into a downstream model. Within this example, we showed you how you reference snapshots and how to query them for the active record or a record at a point in time.

While snapshots are a great tool within dbt, they don't come without their pitfalls, so in the final section of this chapter, we covered the potential issues with snapshots. The main issue with snapshots is that they only capture a record when they run, so if a record has been updated twice since the last run of a snapshot, then only the most recent change will be captured. Depending on the velocity of your data, this may be a non-issue, but if it is an issue, we shared that a different type of history tracking, such as log-based replication, may be advantageous to explore.

We hope that this chapter has left you with the knowledge of when to use, and not to use, snapshots, as well as how to implement, configure, and reference them in other dbt models. In the next chapter, we're going to cover how to take your dbt transformations to the next level by using Jinja and macros.

CHAPTER 6

Jinja, Macros, and Packages

Throughout this book so far, we have covered topics such as models and snapshots which primarily make use of mostly straightforward SQL transformations. However, have you ever found yourself writing repetitive SQL code that you wish you could easily use throughout your transformations without needing to rewrite the same code every time? This is where Jinja in conjunction with dbt comes in and adds extensible programmatic capability to your SQL code.

Jinja is a text-based template engine used in Python to generate markup or source code. Jinja enables you to use variables, filters, and control structures (such as for loops) to generate text files at compile time. Within the context of dbt, Jinja compiles to SQL code. For Data Engineers that write Python, this should feel very familiar; however, for Data Analysts that primarily write SQL, this can be intimidating at first. While there is some advanced stuff you can do with it, we feel like the learning curve is minimal in getting started.

Besides adding programmatic capabilities to your SQL transformations, Jinja also provides you the ability to create macros. Recall from Chapter 1 that Jinja macros are comparable to functions in object-oriented and functional programming languages. Throughout this chapter, we will cover how to use Jinja within models and how to abstract reusable logic into macros. If you have never used Jinja before, this chapter will provide you with the fundamental knowledge you need to get started.

Although it is great to have an understanding of how to build macros yourself, before you start building a solution it's wise to check to see if someone else in the dbt community has run across a similar problem because they probably have! Toward the end of the chapter, we will discuss how to access the dbt package hub and several useful packages that are available.

C. Cyr and D. Dorsey, *Unlocking dbt*, https://doi.org/10.1007/978-1-4842-9703-2_6

Jinja Basics

This section is intended to provide you with the fundamental knowledge that you need to start exploring the world of using Jinja within a dbt project. There are many features of the Jinja template engine that you may be interested in learning about that we do not cover within this section. If this sounds like you, then we would recommend that you check out the official Jinja documentation at `https://jinja.palletsprojects.com`.

Common Syntax

Let's start by just looking at the common syntax of Jinja. Jinja makes use of the curly brace characters, commonly referred to as "curlies" in the dbt community, to define delimiters within a template. There are three default delimiters that you will encounter when working with Jinja in the context of dbt:

- `{{ ... }}` to represent expressions. Expressions compile to a string and are frequently used to reference variables and macros.
- `{% ... %}` to represent statements. Statements are used to define macros, snapshots, and custom materializations. Additionally, statements are used to identify control structures.
- `{# ... #}` to represent comments. Text within the comment delimiter will not compile with the rest of the Jinja template.

As you've worked through this book so far, you have run into Jinja many times, but may not have realized it. For example, the `{{ source() }}` and `{{ ref() }}` functions are both Jinja expressions. dbt makes use of Jinja expressions to dynamically place table references into your SQL code at compile time. In addition to this, you have encountered Jinja statements as well. Recall from the prior chapter that anytime you start a new snapshot, you have to wrap your SQL code in `{% snapshot %} ... {% endsnapshot %}`.

Expressions

Expressions are one of the most fundamental concepts of Jinja, and you will find yourself using them frequently in dbt as you become more comfortable with incorporating Jinja into your SQL code. The most basic explanation of an expression is that at compile time any valid content between the double curlies will be printed as a string in your compiled SQL files. As far as what is considered "valid content," we will cover the following:

- Literals
- Math
- Comparisons

While this isn't a complete list of what you can use within the context of Jinja expressions, we believe that these give you more than enough knowledge to use Jinja within the context of dbt. Let's take a look at a few examples of how expressions in these different categories will compile. While we are including these examples in-line in this section, we have provided you with a .sql file that has all of these examples written out. You can find this file at the following path: *~/models/miscellaneous_chapter_resources/chapter_6/expression_examples.sql.*

If you are familiar with Python, literal expressions should come natural to you because literals are simply Python objects. Some of these objects include strings, numbers, and lists. In the following example, we have expressions that represent these three object types in the respective order:

```
{{ "This compiles to a string" }}
{{ 100 }}
{{ ['this', 'is', 'a', 'list', 'of', 'objects'] }}
```

If you are referencing the example file that we included, you can compile it in one of two ways depending on how you are using dbt. If you are using dbt Core, you will want to run this command to compile these Jinja expressions: `dbt compile --select expression_examples`. You then should be able to find the compiled .sql file in the *target* subdirectory of your dbt project. On the other hand, if you are a dbt Cloud user, all that you need to do is have the example file open and click the compile button. This will present the compiled output in the window below the .sql file in the IDE.

The compiled results look like this:

```
This compiles to a string
100123
['this', 'is', 'a', 'list', 'of', 'objects']
```

In addition to literal expressions, Jinja provides you the ability to use math expressions. We will keep this section brief, but in our experience whenever we need to do math in our transformations, we tend to just use SQL. However, we feel it is important for you to be aware of this functionality in case you have a scenario where it makes sense to use math expressions.

There are many different math operations that you can use in math expressions, but the following example shows you how to add, subtract, and multiply:

```
{{ 19 + 23 }}
{{ 79 - 37 }}
{{ 7 * 6 }}
```

As with all Jinja expressions, these will still compile to a string value. All of these math expressions evaluate to 42.

The last type of expression that we want to cover is comparison expressions. These expressions will return a boolean value of True or False. The comparison operators available for you to use are alike of those available in the Python language. The following are a few examples of comparison expressions:

```
{{ 1 == 1 }}
{{ 1 != 1 }}
{{ 1 < 2 }}
{{ 1 > 2 }}
```

These expressions will evaluate to a respective True or False value based on the appropriate logic. Keep in mind that the output is just a string of "True" or "False." This is an important concept to keep in mind whenever you are working with Jinja expressions. The result is *always* a string.

Before we close this section out, we want to remind you that expressions will very frequently be used in dbt to reference variables or call macros. We showed you earlier in this section how you have been using expressions to call the source and ref macros throughout the examples in this book, but later in this chapter we will show you how you can build and call your own macros or utilize open source macros via packages. We haven't yet shown you how to reference variables using expressions because we first want to introduce you to how to set variables. In the next section, we will show you how to set and reference variables.

Variables

Jinja variables are useful for storing an object for later use or manipulation by setting them using the statement syntax that we discussed earlier. Furthermore, they provide us a way to reference the same object multiple times throughout a template, which in the case of dbt renders as a .sql file. Later in the chapter, we will get into examples of using variables to help us dynamically generate SQL statements, but for now let's discuss how to set and reference variables. You can find file associated with these examples at the following path: *~/my_first_dbt_project/models/miscellaneous_chapter_resources/chapter_6/variables_examples.sql.*

We're first going to look at the first example within the file referenced earlier, also shown in Listing 6-1, to start understanding how variables are defined, how to reference them, and what the referenced variable looks like once it has been compiled.

To set a variable, you will need to use the `set` tag within a statement delimiter. In Listing 6-1, we have shown an example of this by setting the variable `foo` to be equal to the string literal `'bar'`. While we set this variable to be equal to a string, we could have set it to any other type of valid object such as a number, list, or a reference to another variable.

Listing 6-1. Example of setting and referencing a Jinja variable

```
{% set foo = 'bar' %}
{{ foo }}
```

Following the statement where we set the variable `foo`, we have an expression delimiter surrounding our variable. This is how you reference an assigned variable using Jinja. Do keep in mind that if you try to reference the variable before setting it, you will get a compilation error stating that the variable is undefined. However, since we have both defined and referenced the variable, you can run the `dbt compile` command, or if you're using dbt Cloud, just click the compile button.

Once you have compiled this example, all that you should see in the compiled result is the string `bar`. This is fantastic, but what if we wanted to override this variable later on? Let's take a look at the example in Listing 6-2 to see how we can override the assignment of a variable.

Listing 6-2. Example of variable reassignment in Jinja

```
{% set foo = 'bar' %}
{% set foo = 'baz' %}
{{ foo }}
```

Notice that only one adjustment was made from the original code in Listing 6-1 to derive Listing 6-2, which was to add an additional set tag and reassign the `foo` variable to equal `'baz'`. When you compile this example, you should see an output only of the string baz. This sort of variable reassignment becomes very useful when you start to introduce control structures, such as for loops, into your Jinja-fied SQL code.

We cover control structures in the next section, but before we move on, let's cover one final example of setting variables. This time, we are going to use what is known as an assignment block in Jinja. Listing 6-3 provides an example of using an assignment block. This looks very similar to the way we assigned a value to the `foo` variable in the earlier examples except we now have the ability to set more complex variables. In the context of dbt, you will sometimes use this sort of a format to set a variable to a SQL query that you can run.

Listing 6-3. Example of setting a Jinja variable using an assignment block

```
{% set foo %}
'bar'
{% endset %}
{{ foo }}
```

We will discuss more complex and practical examples later in the chapter. Setting a variable in Jinja shouldn't be confused with passing a variable in from the dbt_project.yml or from the command line. We will discuss how this is done later in the chapter, but currently we are just talking about variables for local use in one file. For now, review the compiled results of this example. Notice that the compiled results in Listing 6-1 are different from that of Listing 6-3. In a normal assignment, the quote characters around a string do not show up in the compiled result, but this isn't true for assignment blocks. Assignment blocks, like the one in Listing 6-3, will include all of the characters that you list out. This is very important to keep in mind when you are creating variable assignments because you will need to know whether or not you want the quote characters to show up in your compiled result. If in our example we didn't want the quote characters in our compiled result, we could simply remove them from `bar` in the assignment block.

Conditionals and For Loops

In the prior section, we introduced you to setting and referencing variables with Jinja. The use of variables becomes more advantageous when you combine them with the control structures available in Jinja. The Jinja documentation defines control structures as

> "Control structure refers to all those things that control the flow of a program – conditionals (i.e. if/elif/else), for loops, as well as things like macros and blocks"
>
> https://jinja.palletsprojects.com/en/3.1.x/templates/#list-of-control-structures

While there are several concepts included in this definition, up next we will focus on conditional if statements and for loops. Throughout this section, we will show you how to implement if statements and for loops using Jinja. Both of these are very useful for helping generate repetitive SQL and building macros. In this section, we will focus on the basics of these types of control structures and then provide you with an example of how they can be used to help remove repetitiveness from a SQL select statement. Before we move into examples, let's define both of these:

- **Conditional if statements**: Logical flows that tell a program what to do based on if a condition is true or false
- **For loops**: An iterative process in a program that executes as long as a condition is true

The if statement in Jinja is very similar to that of the Python language. As such, it follows the same if/elif/else structure. Looking at the example in Listing 6-4, you will see how these three different keywords are used. This example can be found at the following path: *~/my_first_dbt_project/models/miscellaneous_chapter_resources/chapter_6/if_statement_example.sql.*

Listing 6-4. Example of a Jinja if statement

```
{% set foo = 'baz' %}

{% if foo == 'bar' %}
    'This condition is checked first.'
{% elif foo == 'baz' %}
    'This is true!'
```

```
{% else %}
    'This is only true if all of the above are false.'
{% endif %}
```

In the preceding example, we build onto examples from earlier in the chapter where we discussed variable assignment. This example starts by setting a variable, `foo`, to the value of `'baz'`. Following this, we defined an if statement with three conditions. In this structure, each of these conditions is checked, in order, until one evaluates as true. For our example, the second condition is true, and as such when you compile this example, the corresponding string, "`This is true!`", will be injected into the compiled .sql file. If control flows, such as if statements, are a new concept to you, then you may want to try changing this example up and making it your own to get a good feel for how to use if statements effectively.

Let's now take a look at the basic syntax of Jinja for loops and how we can use them to help simplify a repetitive SQL statement. For loops are used to iterate over different types of sequences including lists, tuples, and dictionaries. We can then execute a statement for each element in the sequence. In Listing 6-5, you can see that we first initialize the list `foo` with three string elements in it. Having this list initialized allows us to reference it in the for loop.

Listing 6-5. Example of a Jinja for loop

```
{% set foo = ['bar', 'baz', 'qux'] %}

{% for i in foo %}
  Current element in list: {{ i }}
{% endfor %}
```

The for loop follows this syntactic structure:

- Start with a statement block. Here, we tell the Jinja engine to iterate over every value in the `foo` list. The value `i` in this statement is how we can reference the element that we are currently iterating over.
- Next, you define what should be done for each element in the sequence. In this example, we are simply printing a string that tells us what element we are currently operating on.
- Lastly, you must end a for loop with the `{% endfor %}` statement so that the Jinja engine knows where the for loop stops.

Let's expand on for loops and see how we could use them to simplify a SQL statement such as the one seen in Listing 6-6. You can find this example at this path: *~/my_first_dbt_project/models/marts/operations/fct_daily_orders.sql.* To understand what this query is doing, suppose you have been tasked with finding the number of orders placed each day for each status that an order can be in. An order can be in one of three statuses: placed, shipped, or returned. Based on this requirement, you determine that it makes sense to cast the `OrderPlacedTimestamp` to a date and then take the count of the `OrderId` column when it matches the corresponding `OrderStatus` in a case statement.

Listing 6-6. Example of a SQL select statement with a repetitive pattern

```
select
  cast(OrderPlacedTimestamp as date) OrderPlacedDate,
  count(case when OrderStatus = 'placed' then OrderId end) as
  NumberOfOrdersPlaced,
  count(case when OrderStatus = 'shipped' then OrderId end) as
  NumberOfOrdersShipped,
  count(case when OrderStatus = 'returned' then OrderId end) as
  NumberOfOrdersReturned
from {{ ref('fct_orders') }}
group by 1
```

As you can see, the pattern of creating a case statement wrapped in the count function is repeated for each of the three order statuses in our database. You may already be thinking to yourself that we could use a for loop to simplify this process for us. Take a look at Listing 6-7, found in the same file as Listing 6-6, to see how we can alter this query to leverage Jinja.

Listing 6-7. Example of generating repetitive SQL with a Jinja for loop

```
{% set statuses = ['placed', 'shipped', 'returned'] %}

select
  cast(OrderPlacedTimestamp as date) OrderPlacedDate,
  {% for i in statuses %}
  count(case when OrderStatus = '{{ i }}' then OrderId end) as
  NumberOfOrders{{ i }}{% if not loop.last %},{% endif %}
```

```
  {% endfor %}
from {{ ref('fct_orders') }}
group by 1
```

Notice how in this example we only needed to write the count logic once, yet both Listings 6-6 and 6-7 compile to the exact same SQL. Now, for this example, depending on your school of thought, the use of a for loop is maybe overkill since we only have three statuses to iterate over. With that said, it is quite easy to imagine a real-world scenario where that list moves into the 10s, 100s, or even more.

Filters

Let's continue our overview of Jinja by starting to look at template filters. There are bound to be times that you are working with Jinja and need to alter the data within a variable before the template compiles to SQL. Filters take the input of a variable and alter the way it looks or the structure of it.

Jinja filters are applied to variables using the | character. Take a look at the following example to see what it looks like to apply the reverse filter:

```
{% set foo = 'bar' %}
{{ foo | reverse}}
```

Compiled result:

```
rab
```

While this example is very simple, it is effective at showing the basics of Jinja filters. Filters are a broad concept, and we simply want to make you aware of their existence because at times they are useful in the context of dbt. Before we move on, we also want you to be aware that you can chain filters. Let's expand on our prior example and also apply the capitalize filter, which will capitalize the string:

```
{% set foo = 'bar' %}
{{ foo | reverse | capitalize }}
```

Compiled result:

```
Rab
```

There are numerous filters available that are tested and supported by the maintainers of Jinja. You can get a full list of the available built-in filters here: `https://jinja.palletsprojects.com/en/3.1.x/templates/#list-of-builtin-filters`.

Whitespace Control

In this section, we are going to discuss whitespace in Jinja templates as well as how we can control it. Let's start by gaining an understanding of how Jinja applies whitespace in a rendered document, which in our case will always be a .sql file. Instead of trying to explain it, it is likely easier to show you a simple example. Take a look at the following example to see both what the document looks like before and after being compiled:

```
{% set my_var = 'Will there be whitespace here?' %}
{# Will this comment leave whitespace? #}
{% for i in range (1) %}
  'How much whitespace will this loop leave?'
{% endfor %}
```

Compiled result:

```
  'How much whitespace will this loop leave?'
```

As you can see, when the template is compiled, a lot of whitespace is left behind in the rendered document. This is something that was surprising to us when we first started using Jinja. The rule of thumb is that all whitespace outside of a Jinja delimiter will be rendered in the target document. So that means in our example earlier, the first three lines and final line of code leave behind whitespace. That's because before each of those delimiters is a line break!

Fortunately, Jinja provides us with a way to clean up and control whitespace, but be warned that this can quickly become a metaphorical rabbit hole. So, before you go off and start "fixing" all of the whitespace in your dbt project, consider what value is being added.

Now that we have gotten the disclaimer out of the way, let's take a look at how we can actually clean up whitespace with Jinja. To remove whitespace, we will use the minus sign, -, which can be used to remove either leading or trailing whitespace:

- To remove *leading* whitespace, add a minus sign at the beginning of the Jinja delimiter: `{%- ... %}`
- To remove *trailing* whitespace, add a minus sign at the end of the Jinja delimiter: `{% ... -%}`
- Or remove *trailing* and *leading* whitespace by adding a minus sign in both locations: `{%- ... -%}`

Now that we have an understanding of how to clean up the whitespace in rendered documents that contain Jinja, let's look at how we can use whitespace control to clean up our prior example. Suppose that we want our compiled .sql file to look like this:

```
'How much whitespace will this loop leave?'
```

To accomplish this, we need to do two things. The first is to remove the indentation before our string since that is whitespace outside of a Jinja delimiter it will compile to the final file. The second is to use the whitespace control character, -, to remove all of the whitespace associated with our variable assignment, comment, and for loop. This can all be achieved by changing the example to this:

```
{%- set my_var = 'Will there be whitespace here?' %}
{#- Will this comment leave whitespace? #}
{%- for i in range (1) -%}
'How much whitespace will this loop leave?'
{%- endfor %}
```

This wraps up the section of this chapter where we introduced you to the basics of Jinja and how you can start using them in the context of dbt. So far, we really have only shown you how you can use Jinja to help generate repetitive select statements, but up next we are going to introduce you to building macros that can be used across many of your SQL transformations.

Building Macros

Several times throughout this chapter, we have hinted at the idea that macros are a way that you can use Jinja to abstract reusable SQL logic. This is a concept that will feel very familiar if you come from a background in functional programming or have made use of SQL user-defined functions (UDFs) before. There are many practical use cases for incorporating macros into your dbt project ranging from a simple macro that adds two numbers to a set of macros used to minimize downtime during deployments.

Throughout some of the remaining chapters of this book, we will use macros to help us solve problems that are more complex than the examples we are going to show in this section. We recommend using this section as a starting point to begin understanding the syntax associated with macros and the basics of how they work.

Before we start building macros, you should be aware where dbt expects macros to be stored. You may have already recognized the presence of a directory in your dbt project named *macros*. This directory was created when you ran the `dbt init` command to initialize your project, and this is where dbt expects macros to be stored. Macros, like most things related to dbt, are simply .sql files, and within these files we define macros using the macro statement block. We will use the next few sections to show you examples of building macros so that we can provide you with the tools that you will need to build macros of your own.

Phone Number Formatter Macro Example

Let's jump right into our first example and build a macro that accepts a phone number in an initial format, `1-123-456-7890`, and converts it to a different format, `(123) 456-7890`. We will make the assumption that the input will always be in the first format shown earlier, so we aren't going to put any additional logic to check the initial format. With that said, macros are flexible, and if this is something you want to try to incorporate, go for it!

We've already defined this macro for you, and you can find it at this path: *~/my_first_dbt_project/macros/business_logic/generic/phone_number_formatter.sql*. Let's walk through this file to understand what this macro is doing and how we can reference it in a model. The first thing that we want to point out to you is how the actual logic of this macro is surrounded by what we will refer to as the *macro block*. The macro block does three things for us:

1. Tells dbt that this is a macro, and as such every macro you write should be wrapped in this block.

2. Provides a name for the macro.

3. Allows you to define the arguments that the macro accepts.

In our example, the macro block looks like this:

```
{% macro phone_number_formatter(initial_phone_number) %}
...
{% endmacro %}
```

From this example, you can see that we have named this macro phone_number_formatter and defined that it accepts one argument, initial_phone_number. Within the macro block is where the logic is actually defined. For our example, we have written a simple SQL transformation that converts the input into the phone number format mentioned at the beginning of this section. All together, the macro looks like this:

```
{% macro phone_number_formatter(initial_phone_number) %}
  '(' || substr({{ initial_phone_number }}, 3, 3) || ')' || ' ' ||
  substr({{ initial_phone_number }}, 7, 9)
{% endmacro %}
```

Before we move on and look at how to call this macro from a model, we first want to point out how arguments are referenced within a macro. Notice that when we reference the initial_phone_number argument, it is wrapped in double curlies. Recall from our overview of Jinja basics that this is known as an *expression* and is how Jinja knows to convert an argument, or variable, to a string.

Let's now take a look at how to use this macro in a transformation by first navigating to the *stg_crm_customers* model, which can be found at this path: *~/my_first_dbt_project/models/staging/crm/stg_crm_customers.sql*. As you may have guessed, we are going to use this macro to format the PhoneNumber column. To do this, you will need to replace cus.Phone as PhoneNumber in the select statement with {{ phone_number_formatter('cus.Phone') }} as PhoneNumber.

Now that the model is using the macro to format the phone number, you can run the dbt compile command to see what the compiled SQL looks like, or if you are using dbt Cloud, you can simply click the compile button in the UI. As you can see in the following compiled SQL, the Jinja statement calling the macro is replaced with the actual contents of the phone_number_formatter macro:

```
select
...
'(' || substr(cus.Phone, 3, 3) || ')' || ' ' || substr(cus.Phone, 7, 9) as
PhoneNumber
...
```

Note If you find yourself coding repetitive transformations in your models (like formatting a phone number), then you should consider utilizing or creating a macro to handle that. As always, first check to see if one already exists before building new.

Return a List from a Macro Example

The example we just walked through provided you a basic understanding of how to put reusable SQL into a macro and then call that macro from within a model. As you work more with dbt, you will run across scenarios where this is useful, but reusable SQL isn't the only reason to build macros. There are times when it may be useful to have the macro *return* a value, list, or some other type of object for you to work on within a model. Recall from earlier in the chapter, we provide you an example, in Listing 6-7, of how to generate SQL using a for loop. In that for loop, we iterate over a list of order statuses.

In this example, we are going to build off of that same example, but instead of defining the list of order statuses within the same file, we are going to move the list to a macro. Let's define the macro like this:

```
{% macro get_order_statuses() %}
  {{ return(['placed', 'shipped', 'returned']) }}
{% endmacro %}
```

Within this macro, we are doing nothing more than moving the list of order statuses here to make that list callable from anywhere within the dbt project. To do so, we are using `return` which is a Jinja function built into dbt. The use of the return function is the first time we have introduced you to any of the built-in dbt Jinja functions, and later in the chapter, we will further explore more of these commonly used functions.

Now that this list has been placed in a macro, we can consistently call this list from anywhere in our dbt project, and if the list ever needs to be updated, we only need to update it in one place. We have demonstrated how to call this macro in Listing 6-8. This example looks very similar to Listing 6-7, but we no longer need to define the list and instead can call the macro to retrieve the list directly within the for loop.

Listing 6-8. Example of calling the get_order_statuses macro that returns a list

```
select
  cast(OrderPlacedTimestamp as date) OrderPlacedDate,
  {% for i in get_order_statuses() %}
  count(case when OrderStatus = '{{ i }}' then OrderId end) as
  NumberOfOrders{{ i }}{% if not loop.last %},{% endif %}
  {% endfor %}
from {{ ref('fct_orders') }}
group by 1
```

Generate Schema Name Macro Example

So far, we have shown you how to use macros to abstract repetitive SQL and also how you can use the `return` function to return some data to wherever you call the macro from. For the final example of this section, we want to show you how you can override built-in dbt macros to change the behavior of your dbt project. At this point, you should at least be familiar with the `ref` and `source` built-in macros because we discussed these in both Chapters 4 and 5, and while you can override those, we will focus on a different macro in this section.

Recall from Chapter 2 when we walked you through how to connect dbt to your data platform that we showed you how to set the schema for your development environment. Depending on whether you are using dbt Core or dbt Cloud will impact where you set your target schema, but as a quick reminder:

1. For dbt Cloud users, this was set in the credentials section of the profile settings tab in the dbt Cloud UI, as seen in Figure 6-1.

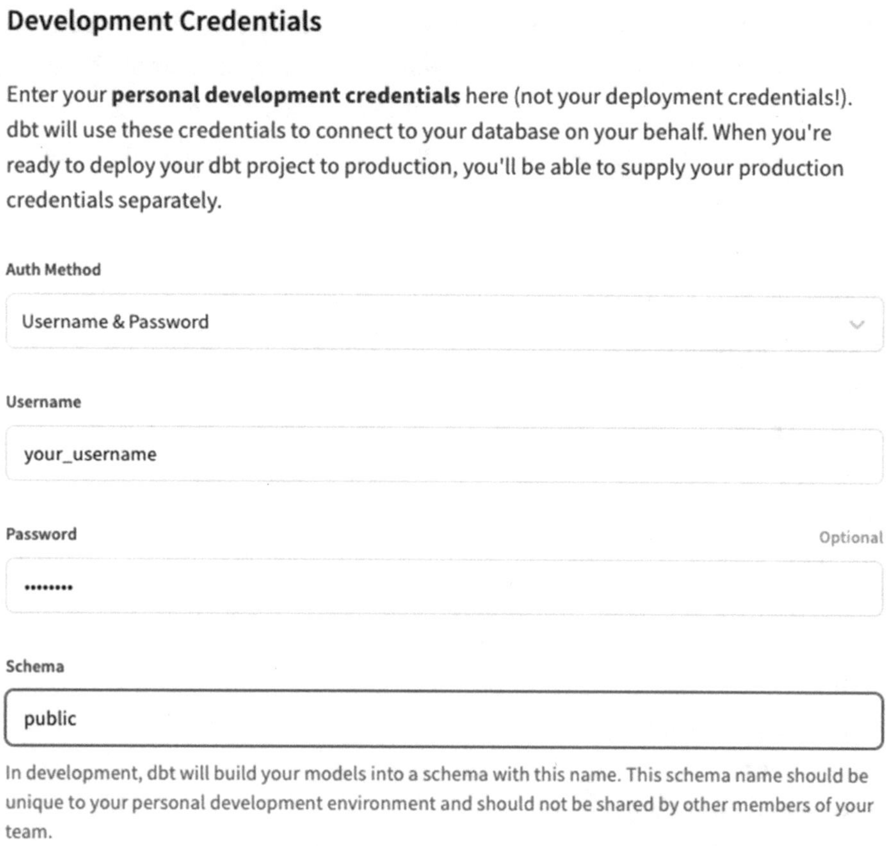

Development Credentials

Enter your **personal development credentials** here (not your deployment credentials!). dbt will use these credentials to connect to your database on your behalf. When you're ready to deploy your dbt project to production, you'll be able to supply your production credentials separately.

Auth Method

Username & Password

Username

your_username

Password Optional

••••••••

Schema

public

In development, dbt will build your models into a schema with this name. This schema name should be unique to your personal development environment and should not be shared by other members of your team.

Figure 6-1. *Where to set the development schema in the dbt Cloud UI*

2. For dbt Core users, this was set in the *profiles.yml* under the schema configuration as follows:

```
my_first_dbt_project:
  target: dev
  outputs:
    dev:
      ...
      schema: public
      ...
```

With that in mind, as you built the example models from Chapter 4 recall that we set custom schemas for several of these models. You can see an example how these configurations were set in the *dbt_project.yml* below. For example, we configured that all models in the *staging* directory should be compiled into a schema named staging:

```
models:
  my_first_dbt_project:
    staging:
      +schema: staging
      crm:
      furniture_mart:
    intermediate:
      +schema: staging
      finance:
      operations:
    marts:
      finance:
        +schema: finance
      operations:
        +schema: operations
```

However, you may have noticed that when dbt generates these custom schemas, it does it in a quite interesting way. Instead of just generating models in the custom schema that you define, it generates the schema name as *<target schema>_<custom schema>*. So going back to our staging models' custom schema, dbt will actually render these models in a schema called *public_staging*.

Note This will vary depending on the target schema that you define. For our purposes, we configured the target schema with the name *public*.

Typically, in a dbt development environment, each engineer will define their target schema to be their name, or something similar that is unique. As such, each developer would have a set of custom schemas prepended with their name (or whatever their target schema is). This behavior is ideal in a development environment because it allows numerous engineers to work in their own schemas without having to worry about overwriting each other's work. However, it creates messy schema names in production, but fortunately this behavior can be overridden. All of this default behavior comes from a macro called *generate_schema_name,* and we can override the way this macro works by defining a macro in our project with the same name. Of course, it might be useful to see the default, unedited, code for this included macro before making any edits to it. You can check out the default as follows:

```
{% macro generate_schema_name(custom_schema_name, node) -%}
    {%- set default_schema = target.schema -%}
    {%- if custom_schema_name is none -%}
        {{ default_schema }}
    {%- else -%}
        {{ default_schema }}_{{ custom_schema_name | trim }}
    {%- endif -%}
{%- endmacro %}
```

Note You can find this macro at this path to follow along: *my_first_dbt_project/ my_first_dbt_project/macros/configuration/generate_schema_name.sql.*

For this example, suppose that you have two environments, *dev* and *prod,* and you want to generate schema names differently in each environment in this format:

- In production, render custom schema names as the configured custom schema name. Do not prepend the target schema.
- In all other environments, ignore the custom schema name and materialize everything in the target schema name.

If you follow this pattern, you will be able to clean up the schema names in production while still maintaining a dedicated space for each developer to work in. To do this, you will want to create a .sql file in the *macros* subdirectory. We have already generated this for you, but keep in mind that the name of this file could be anything. Additionally, it is common practice to keep macros one-to-one with files and match the file name to the macro name. This makes for very easy housekeeping when it comes to understanding what is stored within a macro file. However, you can have multiple macros stored within a single file, but if you choose to go with this approach, we caution you to have a standardized naming convention and organizational structure, or you may end up with very messy macro files.

Next, within this .sql file, you will need to define a macro name *generate_schema_name*. Unlike the name of the .sql file, this macro **must** be named exactly like this because when your project is compiled, dbt searches for any macros that override built-in dbt macros. To create the environment-aware schema behavior that we described earlier, you should create the *generate_schema_name* to look like Listing 6-9.

Listing 6-9. Example of a macro to override the default custom schema behavior

```
{% macro generate_schema_name(custom_schema_name, node) -%}
  {%- set default_schema = target.schema -%}
  {%- if target.name == 'prod' and custom_schema_name is not none -%}
    {{ custom_schema_name | trim }}
  {%- else -%}
    {{ default_schema }}
  {%- endif -%}
{% endmacro %}
```

This example is relatively simple if you break down what is happening. The first thing that we are doing, as seen in the following example, is defining the variable `default_schema` to be set to the target schema for the current environment. We will go into more detail on the `target` function in the next section.

```
{%- set default_schema = target.schema -%}
```

Next, an if statement begins where the first set of conditions being checked is to see if the target name is *prod* and see if a custom schema has been defined. Again, we will discuss the `target` function further in the next section, but for now just be aware the `target.name` represents which profile is currently configured to the target:

```
{%- if target.name == 'prod' and custom_schema_name is not none -%}
```

For now, we have been running all of our dbt transformations in a development environment, so this will always resolve to be false. Later in the book, we will discuss running dbt in production, and at that point we will see in action how this macro affects schema names in production. However, if we were running this in production currently, this macro would satisfy our requirements of excluding the target schema name from being prepended to the custom schema. Instead, only the custom schema name will be used, and whitespace will be trimmed to ensure we have no whitespace in the schema name. This is done by setting the return value to:

```
{{ custom_schema_name | trim }}
```

Lastly, in this if statement for all other conditions, the default schema will be returned by setting

```
{%- else -%}
  {{ default_schema }
{%- endif -%}
```

This block would satisfy the second requirement of all models materializing in the developers target schema regardless of whether or not a custom schema was set.

This is just a starting place for how you can override dbt's default behavior for custom schemas, and we encourage you to tweak this logic to suit the needs of your environment setup. For example, it is common for teams to have a QA environment, and it may be worthwhile for that environment to more similarly reflect the production environment. However, in the current state of this macro, the QA environment would instead look similar to the development environment.

Before we move on, we wanted to call out that if you want to have dbt build your schemas using the logic that we described earlier, you could simplify the macro to simply be

```
{% macro generate_schema_name(custom_schema_name, node) -%}
  {{ generate_schema_name_for_env(custom_schema_name, node) }}
{%- endmacro %}
```

Here, instead of defining the logic as laid out in Listing 6-9, we call another built-in macro named *generate_schema_name_for_env*. This works because the logic in this built-in macro is exactly the same as what we defined in Listing 6-9. While you can do this, we typically haven't on projects that we've worked on because it adds another layer of complexity if you need to change the behavior of the *generate_schema_name_for_env* macro. Not to say you can't do it, but you should consider the trade-offs before selecting an approach to how you want to handle custom schema names.

dbt-Specific Jinja Functions

So far, we've shown you the basic Jinja skills you will need to work with templated SQL and introduced you to different use cases for building and using macros. To take both of these learnings to the next level, we want to introduce you to some of the Jinja functions that are available to you via dbt. In the next few sections, we are going to walk you through several of these functions that we have found to be very useful as we have built dbt projects in the past. However, within the scope of this book, we simply cannot cover every Jinja function available for you to use in your own projects, so we would recommend reviewing dbt's documentation to see the full list of available functions.

Note You can access the full list of dbt Jinja functions from dbt's documentation here: *https://docs.getdbt.com/reference/dbt-jinja-functions*.

Target

We introduced you to the target function in Listing 6-9 in the prior section when we showed you how you can use macros to change the behavior of how dbt generates custom schema names. However, we relied on you trusting that it would work, but we didn't provide much detail about the target function or how it works.

The target function allows you to access information about the connection to your data platform. You can access different attributes of the target such as the name, profile name, or schema. Recall that the target is the connection information that dbt needs to connect to your data platform. In our experience, the target function is most commonly used to change the behavior of your dbt project based on what environment your project is currently running in. We will show you an example of this later, but for now let's look at which attributes are available to you and how they are accessed.

There are currently five attributes of the target that are accessible by all dbt users, regardless of which adapter is being used to connect to your data platform:

- `target.profile_name` represents the name of the profile actively being used to connect to your data platform. If you aren't sure what this name is, you can find it in both your profiles.yml and dbt_project.yml.
- `target.name` represents the name of the target. While not always true, the target name tends to be synonymous with an environment name.
- `target.schema` represents the name of the schema where dbt will generate models in your database.

- `target.type` represents the adapter being used to connect to your data platform. For our examples, this will be *snowflake*, but could also be *bigquery, databricks, etc.*
- `target.threads` represents the number of threads that you have configured dbt to use.

As mentioned earlier, these five attributes are available regardless of which adapter you are using. For some adapters, there are additional attributes that are available that you may find useful. Instead of going over each adapter, we are going to focus on the universally available attributes, but if you are interested in learning about how you can use the target function attributes that are specific to the adapter you're using, we will refer you to dbt's documentation.

In the following example, we show you how you can change the materialization type for an individual model based on the target name:

```
{{
  config(
    materialized = 'view' if target.name == 'dev' else 'table'
  )
}}
```

Here, we use `target.name` to check the target name and materialize this model as a view if the target name is `'dev'`, but for all other target names, this model should be materialized as a table.

This

The `this` function is useful for accessing the full reference to your model as it would appear in your database, meaning that it will compile as a string in the format `database.schema.table`. You can also access each of these three attributes individually by calling

- `{{ this.database }}`
- `{{ this.schema }}`
- `{{ this.identifier }}`

You may be wondering why we don't simply use the ref function to reference the current model's database identifier, but we can't because dbt will throw an error due to the presence of a circular dependency. So instead, we can use the this function as a workaround.

Note A circular dependency is when a model references itself. Accidental circular dependencies should be avoided, but when they are intentional (such as in incremental logic), that can be valuable to use.

As we discussed in Chapter 4, we can use the this function as a means to build models incrementally. Take a look at the following example to see how we can use this function to limit the records that are being pulled into an incremental model run by filtering on a timestamp:

```
{{
  config(
    materialized = 'incremental'
  )
}}
select
    *
from {{ ref('foo') }}
{% if is_incremental() %}
  where updated_at >= (select max(updated_at) from {{ this }})
{% endif %}
```

While the preceding example is a very common use case for the this function, it isn't the only thing that it is useful for. You can also use this function to build logging or call the model in a hook. For example, you can use this function in a post-hook to operate on the resulting materialized table or view in some way, such as handling hard deletes. We will discuss hooks and how to use them in the next chapter.

Log

The log function can be used to write a message to the log file whenever a dbt command is executed. Optionally, you can also have the message written to the terminal. We have often found this function useful for messages that are used for debugging. The following example shows you how you could generate a log message that will write the current invocation id for the dbt command that was executed to the log file and print the message to the terminal. This statement could be altered to not print to the terminal by removing the info argument or setting it to False.

```
{{ log('invocation id: ' ~ invocation_id, info=True) }}
```

Tip In Jinja syntax, the tilde (~) is used for string concatenation.

Adapter

Throughout this book, we have referenced the term *adapter* quite regularly; under the hood, an adapter is an object that dbt uses to communicate back and forth with your database. The adapter Jinja function is an abstraction of the adapter object which allows you to invoke different methods of the adapter class as SQL statements. There are many different methods that you could invoke, but we are going to show just one example where we will get a list of columns in an existing relation. For a more comprehensive list of adapter methods, we will refer you to the official dbt documentation.

The following example shows how to use the get_columns_in_relation method to return a list of columns in the *fct_orders* table. Once you have this list of columns available, you can operate on the list in any capacity that you desire.

Note This method returns a list of Column objects. Each object has several attributes including column, dtype, char_size, numeric_size, and more.

```
{% set columns = adapter.get_columns_in_relation(ref('fct_orders')) %}
{% for column in columns %}
     {{ log('Column object: ' ~ column, info=True) }}
{% endfor %}
```

If you either run or compile this example, you will see several rows printed to the terminal similar to

```
Column object: SnowflakeColumn(column='ORDERID', dtype='NUMBER',
char_size=None, numeric_precision=38, numeric_scale=0)
```

While it is useful to be able to access all of this metadata using the Column object, for our purpose we simply want a list of column names. We can alter the initial example to achieve exactly this:

```
{% set column_objects = adapter.get_columns_in_relation(ref('fct_
orders')) %}
{% set columns = [] %}
{% for column in column_objects %}
    {% do columns.append(column.column) %}
{% endfor %}
{% for column in columns %}
    {{ log("Column: " ~ column, info=True) }}
{% endfor %}
```

If you now run or compile this modified example, you should see each column name printed to the terminal similar to this:

```
Column: ORDERID
...
Column: ORDERPLACEDTIMESTAMP
```

Var

The var function is used to access the value of a variable that was set in a project's *dbt_project.yml* file. This is useful for setting variables that are used in many places throughout a project. This aids in the maintainability of your project because if the value of this variable needs to be changed, it can be updated in your *dbt_project.yml*, and the change will trickle down throughout your project.

Fortunately, the var function is flexible, and it can be used in both .sql files and .yml files. In the following example, we are going to show you how you can use this function in a *sources.yml* file when defining the database that a source references. We didn't include this example in our example code repo because we aren't using sources, but you could easily apply this example in a real-life scenario.

We first need to define the variable in the *dbt_project.yml* as follows:

```
...
vars:
  raw_db: raw
...
```

Suppose that you have a *sources.yml* file that looks like this:

```
version: 2
sources:
  - name: raw
    database: raw
    schema: public
    tables:
      - name: orders
...
```

To swap out the static database reference in the *sources.yml*, we will replace the database config like this:

```
version: 2
sources:
  - name: raw
    database: "{{ var('raw_db') }}"
    schema: public
    tables:
      - name: orders
...
```

The prior example required you to declare the variable in the *dbt_project.yml*, but you can actually reference a variable within a model that hasn't been declared. However, in the next example, we will want to make use of the second, optional, argument of the var function - default. This second argument allows you to set a default value if the variable hasn't been declared. Essentially, this ensures that a value is always substituted in for the variable when dbt compiles your project.

In the following example, we are going to use the var function inside of an incremental model so that we can run backfill jobs without having to fully refresh the entire table. As a quick refresher from Chapter 4, incremental is a model materialization strategy that is used to merge records into the destination table instead of rebuilding the table each time dbt is run. It is common practice to use the is_incremental function to add a where clause to incremental runs of a model to limit the dataset, oftentimes based on a timestamp field, to improve performance.

However, when building tables incrementally in a data warehouse, it is common to need to backfill data due to adding a new column, changing logic, or some other number of reasons. dbt provides the --full-refresh flag that can be used to fully refresh incremental models, but this can be computationally expensive. As an alternative to fully refreshing models, the following example will enable you to pass a date in as a command-line argument at run time to run a backfill on a model up to a certain date. We will use the default argument of the var function so that if no date is passed as a command-line argument, the date to base the incremental logic on can just be the maximum date from the model itself.

Let's start with a simple example of an incremental model. This example should look familiar because it is the same one we used as an example for the this function. Take note that in the is_incremental block, we are using a subquery to retrieve the max updated_at timestamp from the destination table and filtering the results from the select statement based on that timestamp.

```
{{
  config(
    materialized = 'incremental'
  )
}}
select
    *
from {{ ref('foo') }}
{% if is_incremental() %}
  where updated_at >= (select max(updated_at) from {{ this }})
{% endif %}
```

The code in the preceding example is great for incrementally loading the destination table on a schedule, but the time will come when something about the schema changes or there is a data quality issue that needs to be dealt with. In this scenario, the easiest option would be to do a full refresh, but what if you have restraints, such as performance or cost, that limit you from doing this? That is where we can take advantage of the following modified example, where we use the max updated_at timestamp as a variable's default value, but also provide the ability to pass in a date at run time:

```
{{
  config(
    materialized = 'incremental'
  )
}}

{% if is_incremental() %}
  {% if execute %}
    {% set query %}
      select
        max(updated_at)
      from {{ this }}
    {% endset %}
  {% endif %}
  {% set max_updated_at = run_query(query).columns[0][0] %}
{% endif %}

select
  *
from {{ ref('foo') }}
{% if is_incremental() %}
  where updated_at >= '{{ var("backfill_date", max_updated_at) }}'
{% endif %}
```

Between the initial and modified examples, two changes were made that we want to break down for you. The first is the Jinja that was added to the top of the query, as seen in the following example. This is using a query to select the max updated_at timestamp from `this` model. We first set a variable called `query` that is run using the `run_query` Jinja function. The `run_query` function returns an *agate table*, which is why we have to index the returned agate table to actually get the timestamp value that we want.

Note If you are interested in learning more about Agate, you can read the documentation at the following link, but we don't recommend spending too much time on this topic:

https://agate.readthedocs.io/en/latest/api/table.html

```
{% if execute %}
  {% set query %}
    select
      max(updated_at)
    from {{ this }}
  {% endset %}
{% endif %}
{% set max_updated_at = run_query(query).columns[0][0] %}
```

Next, notice that we also altered the where clause that will be added to the query on incremental runs by using the var function instead of a subquery as we were in the original example. The value for the variable named `backfill_date` will be placed in the where clause if it is supplied; otherwise, the `max_updated_at` timestamp will be used by default. In practice, we would only ever supply the `backfill_date` variable when we want to do backfills. Let's now take a look at the commands we could run to make this work for us and what the compiled SQL would look like as well.

We can run the example that we just reviewed using this command:

```
dbt run --select fct_foo
```

This command would simply run the model with default incremental logic that we laid out. As such, the max updated at timestamp will be pulled from the destination table and be plugged into the where clause. But, let's now take a look at a command that we could run to conduct a backfill to a certain point in time. Suppose that we want to reprocess records with a date that is on or after 2023-01-01; we can use this command to do so:

```
dbt run --select fct_foo --vars {"backfill_date":"2023-01-01"}
```

This command is almost the same as the first one, but it includes the command-line argument `--vars`. This command allows you to pass key-value pairs in as a YAML dictionary, where the *key* is the variable name and the *value* is what you want to have the variable set to. As such, the preceding command would generate the following compiled SQL:

```
select
  *
from {{ ref('foo') }}
where updated_at >= '2023-01-01'
```

Note This will only be the compiled SQL for incremental runs. If this was an initial run of the model, the where clause would be excluded.

Env Var

The final dbt Jinja function that we are going to share with you is the env_var function, which can be used to access environment variables within dbt. Environment variables are useful for dynamically managing the configuration of your dbt project across different environments, removing secure credentials from your code, and adding custom metadata to dbt artifact files.

We will discuss how to set environment variables in dbt Cloud. However, for dbt Core users, we operate under the assumption that you have already set any needed environment variables on your machine. There are simply too many ways that environment variables can be set on a machine, and it varies too much across different operating systems for us to discuss each method.

Tip The `env_var` function can be used everywhere that Jinja can be used, including the *profiles.yml*, *sources.yml*, *.sql* files, etc.

For users of dbt Cloud, any environment variables you create need to be prefixed with DBT_. You can view existing environment variables by navigating to the Environments page and clicking the Environment Variables tab. If you need to set a new environment variable, you will need to click the Add Variable button. In Figure 6-2 is an example of what this page looks like with an environment variable present.

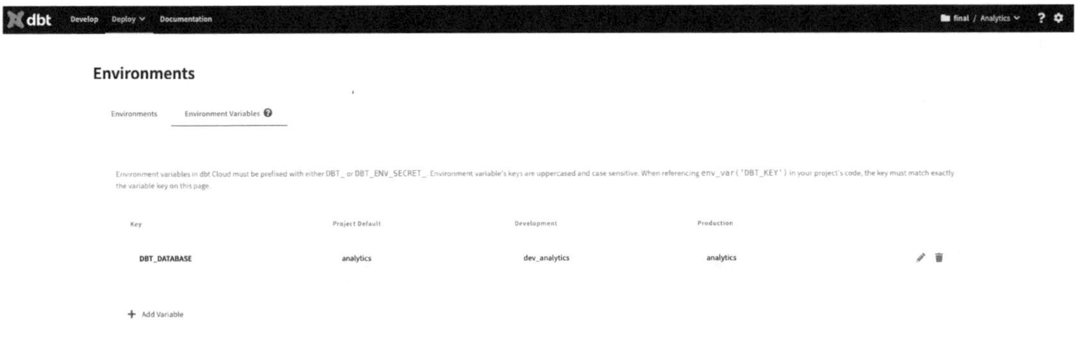

Figure 6-2. *dbt Cloud Environment Variables page*

If you are using dbt Core, there isn't a hard requirement to prefix your environment variables with DBT_, but we recommend doing so, and we will follow this pattern for the remainder of the section. We recommend following this same pattern for these reasons:

- It helps make it clear which environment variables you set on your machine are meant to be used for your dbt project.
- It will be easier to migrate from dbt Core to dbt Cloud if this is ever something you should choose to do.
- Easily control which files can access environment secrets.

Recall from earlier in the chapter that we showed you how you could use the var function to plug in the database name for data sources in the *sources.yml* file. This would look something like this:

```
version: 2
sources:
  - name: raw
    database: "{{ var('raw_db') }}"
    schema: public
```

```
    tables:
      - name: orders
...
```

This is useful if you use the same raw database in all environments, but what if instead you need to use two raw databases to build your dbt models?

- `raw`: To be used in the production environment
- `raw_dev`: To be used in the development environment

This is an opportunity to access environment variables using the `env_var` function. Assuming that you have an environment variable named `DBT_RAW_DB`, set with the appropriate value, you can alter the *sources.yml* to use the environment variable instead of the project variable. Here is what the updated configuration would look like:

```
version: 2
sources:
  - name: raw
    database: "{{ env_var('DBT_RAW_DB) }}"
    schema: public
    tables:
      - name: orders
...
```

While the prior example is useful for dynamically configuring your project based on the current environment, the `env_var` function is also commonly used to ensure that secrets are not hardcoded into your dbt project. Some common secrets that you may want to pull into environment variables include

- Username
- Passwords
- Deployment tokens

If you are using dbt Core and set up your project following the instructions in Chapter 1, it is likely that you have your database credentials stored directly in your *profiles.yml*. This likely isn't a huge problem in your development environment, but nonetheless you could make updates so that your username and password are stored as environment variables instead. To do so, first create two environment variables named DBT_ENV_SECRET_USER and DBT_ENV_SECRET_PASSWORD with their respective values. Then update the user and password configs in your *profiles.yml* to

```
your_project_name:
  target: dev
  outputs:
    dev:
      ...
      user: "{{ env_var('DBT_ENV_SECRET_USER') }}"
      password: "{{ env_var('DBT_ENV_SECRET_PASSWORD') }}"
      ...
```

Notice that both of these environment variables were named using the prefix DBT_ENV_SECRET_. While this prefix isn't required, it is ideal for environment secrets because dbt will ensure that all environment variables using this prefix are only accessible by the *profiles.yml* and *packages.yml* files. This helps ensure that your environment secrets aren't accidentally displayed in your database or dbt artifacts (such as the manifest.json file).

Useful dbt Packages

While it is extremely important to have an understanding of how to incorporate Jinja and macros into your dbt project, if you find yourself needing to build complex processes using them we would recommend that you check to see if someone else in the dbt community has already solved the same problem. In this section, we will be introducing you to the dbt package hub, a collection of different dbt packages that are available for you to incorporate into your project.

Within the dbt package hub, you will find projects published by dbt and community members. The packages available range from a collection of common utilities to transformations to logging. In the remainder of this section, we are going to share a few details around a handful of available packages. While there are packages available that solve more niche problems, in our experience the following packages have proven themselves useful for a wide range of project types. As such, we will focus on this narrow subset, but keep in mind that there are many more packages available to you.

Lastly, since packages aren't a part of dbt directly and can be maintained by the community, we will keep this discussion fairly high level. Instead of sharing individual examples, we are more interested in providing you with awareness of what each of these packages can do for you. We then hope that you can determine which packages would be useful for adding functionality to your own project. Additionally, the dbt Slack community and dbt Discourse are invaluable resources to learn about how current dbt users are implementing macros and packages.

Add a Package to Your Project

Before we share any of the available packages, we'll take a brief moment to show you how to add any package to your project. First and foremost, the list of packages that you want to add to your project will need to be defined in the *packages.yml* file.

Note The *packages.yml* file should be in the same directory as the *dbt_project.yml.*

There are several methods that you can use to define which packages you want to add to your project, but for the purposes of this book, we will show you examples of only two of the methods, including

- Adding packages from the package hub
- Adding packages using a git URL

To add a package from the package hub, you use the `package` and `version` configurations in the *packages.yml* file. For example, if you want to add the dbt Utils package and the Codegen package to your project, your *packages.yml* should look like this:

```
packages:
  - package: dbt-labs/dbt_utils
    version: [">=1.0.0", "<1.1.0"]
  - package: dbt-labs/codegen
    version: [">=0.9.0", "<1.0.0"]
```

Take note that we have defined each version as a range instead of a specific version. This allows us to automatically pull any patch updates. However, if you prefer instead to list a specific version, you can.

You can also pull packages using git URLs. This is useful if you have an internal package you want to pull into your project, or if there is a package available on GitHub that isn't available on the package hub. For this example, we will show you how you could pull dbt Utils and Codegen using their git URLs, but keep in mind that this is not the recommended method for packages that are available on the dbt package hub:

```
packages:
  - git: "https://github.com/dbt-labs/dbt-utils.git"
    revision: 1.0.0
  - git: "https://github.com/dbt-labs/dbt-codegen"
    revision: 0.9.0
```

Notice that in the preceding example, we instead use the `git` and `revision` configurations to pull these packages. The `git` config is simply the URL to a git repository, and the `revision` config represents either a tag or a branch name. Also, notice that we had to specify a revision because you can't define a range when pulling packages directly from git.

dbt Utils

The dbt Utils package is one of the most commonly used dbt packages out there. This package contains many useful macros for SQL generation, generic tests, and other helper functions. The macros in this package are designed to work across many different adapters. As such, using macros from this package can simplify your code and make it portable between different dialects of SQL.

In our experience, some of the most commonly used macros from this package include

- `generate_surrogate_key`: Adds a new column to your model to serve as a warehouse surrogate key. This macro accepts a list of one or more columns and returns an MD5 hash.
- `date_spine`: Provides the SQL needed to generate a date spine. This macro is useful for generating a calendar table or modeling subscription type data where you need one row per day for a range of dates.

- `pivot`: A simplified way to pivot row values into columns.
- `union_relations`: Useful for unioning similar, but not equal, relations. For example, you can use this macro to union two tables that have different columns or columns that aren't in the same order.

This, of course, is not an exhaustive list, but if you are interested in learning more about this package and what other macros are available for your use, we recommend you review the package's documentation.

Codegen

As you start working more with dbt, you will quickly learn that YAML is both a curse and a blessing. It is a blessing in the sense that YAML is how dbt can manage the relationships and lineage between your models and is used to generate documentation. But, the curse is that YAML doesn't write itself, and it can be time consuming to do so from scratch.

Fortunately, this is where the Codegen package becomes useful. There are a few light macros in this package that you can execute from the command line using the `dbt run-operation` command. The following are a few of the macros available in this package:

- `generate_source`: This will generate YAML, based on the arguments that you provide, that can be copied in a *sources.yml* file.
- `generate_base_model`: This will generate SQL for you that can be used for base/staging models.
- `generate_model_yaml`: This will generate YAML, based on the arguments that you provide, that can be copied in any *.yml* file where you document and test your models.

dbt Project Evaluator

This package is used to evaluate how well your project is following best practices, as defined by dbt. The package will parse different components to determine

- Model best practices
- Test coverage

- Documentation coverage
- File and naming structure
- Performance

As mentioned, out of the box this package will evaluate your project against dbt Labs' best practices for dbt projects. While dbt Labs has solid opinions and recommendations for best practices, their rules may not work for every project or organization. The developers of the package were cognizant of this and allow you to override variables associated with checks and allow you to disable checks entirely.

Additionally, while it is useful to have this package available to be run in a local development environment, it is even more useful to run this as part of your Continuous Integration (CI) pipeline. It is very straightforward to incorporate this package as a pass/fail check in your CI pipeline, and we will show you how to do so in Chapter 10 when we discuss running dbt in production.

dbt Artifacts

The first three packages that we've discussed are all built and maintained by dbt Labs, but the final two packages that we will share with you were built and are maintained by members of the dbt community.

This package that is maintained by Brooklyn Data will parse the nodes in the graph of your dbt project and build fact and dimension tables in a data mart in your warehouse. The generated models include dimensional data about models, sources, seeds, etc. as well as fact, or event, data such as invocations.

This package is very handy for monitoring the overall health of your project, determining test failure rates, and any other analysis that you can dream up with the available data. Currently, the models that will be included in the generated data mart are

- dim_dbt__current_models
- dim_dbt__exposures
- dim_dbt__models
- dim_dbt__seeds
- dim_dbt__snapshots
- dim_dbt__sources

- dim_dbt__tests
- fct_dbt__invocations
- fct_dbt__model_executions
- fct_dbt__seed_executions
- fct_dbt__snapshot_executions
- fct_dbt__test_executions

dbt Expectations

The final package that we want to share with you is the dbt Expectations package, which contains many generic data quality tests. This package is largely inspired by the Python package Great Expectations.

By utilizing this package, you gain access to data quality test in categories such as

- Range checks
- String matching checks
- Robust table shape checks

We always mention this package to users of dbt because we believe that data quality is one of the most important aspects of data and analytics engineering. More so, this package makes it extremely easy to add robust generic tests to your project without needing to worry about the complexity of building the generic tests. While this package won't have a test for every use case, we recommend always checking here before you start building your own generic test.

Summary

This chapter covered a wide array of topics related to Jinja, macros, and dbt packages. We started this chapter with an overview of the basic Jinja skills that will enable you to be effective with incorporating Jinja into your dbt projects. While we didn't cover every in-depth aspect of Jinja, we did cover the basic syntax and structures, such as variables, control flows, and loops.

Additionally, we provided you with a few examples of how to actually use Jinja to make your SQL code more DRY (Don't Repeat Yourself). However, a word of caution from us is to use Jinja effectively, but sparingly. It is tempting to use Jinja everywhere in your SQL code, but always ask yourself: "Will this simplify my code or make it more complicated for the next developer?" If it doesn't make it more simple for the next developer, you should think twice before doing it.

After covering the basics of Jinja, we offered a few relatively simple examples of how you can abstract repeatable logic into macros. Recall that macros in dbt are synonymous with functions in other programming languages. Although most of the examples we covered are fairly elementary, this section should have provided you with the foundational knowledge needed to start developing macros for your potentially more complex use cases.

As we moved into the next section, we shared with you the Jinja functions that are built into dbt. These functions are useful to call directly within your model files and macros that you have built. Some of the functions that we covered include `target`, `this`, and `env_var`. The list of built-in functions that we shared with you are the functions that we have found to be commonly useful in our experience when developing using dbt. However, there are many more built-in Jinja functions that you can find in the dbt documentation.

We concluded the chapter by introducing you to the concept of importing dbt packages into your project. Packages are simply dbt projects that have been made available to the community to use in projects of their own. We firmly believe that you should see if there is a package available to help you solve a problem you have before you build your own custom macros. We briefly discussed a few packages maintained by both dbt and the dbt community. We intentionally didn't go into a ton of detail on any of these packages because they tend to change more frequently than the fundamentals of dbt Core. As such, we don't want you to be reading outdated information, so if you are intrigued by any of these packages, we recommend that you read their respective up-to-date documentation.

In the next chapter, we are going to discuss hooks. This includes running processes at the beginning or end of a dbt invocation. Additionally, we will also show you how you can run a hook before or after models.

CHAPTER 7

Hooks

SQL is useful for building transformations of raw data into useful models to be used by your downstream consumers, but there are bound to be times when you need to run ad hoc SQL that dbt doesn't support out of the box. For example, you might need to

- Manage the size or capacity of your compute layer
- Apply masking policies or row access policies
- Manage database parameters

This is where hooks can be useful to you for assisting with the management of administrative tasks. Unlike many resources in your dbt project, hooks can utilize SQL commands beyond the simple SELECT statement which opens a world of possibilities. Throughout this chapter, we will refer to all of the discussed operations as *hooks*, but understand that we are referring to three categories of operations defined as

- **Pre-/post-hooks**: Used to run a SQL query before/after the execution of certain dbt nodes.
- **On-run-start/end**: Used to run a SQL query before/after the execution of certain dbt commands.
- **Run operation**: A dbt command used to run a macro. This command was briefly mentioned in the "Codegen" section of Chapter 6, but we will go into greater detail in this chapter.

This chapter will serve as an introduction to how hooks work through both descriptions and examples. Throughout this chapter, you will gain an understanding of how hooks can add flexibility to how your dbt project runs. However, with that in mind, we will also discuss best practices regarding when code belongs in a hook vs. a model.

C. Cyr and D. Dorsey, *Unlocking dbt*, https://doi.org/10.1007/978-1-4842-9703-2_7

Pre-hooks and Post-hooks

Pre-hooks and post-hooks are dbt configurations that can be used to deal with boilerplate code or database administration tasks that dbt doesn't handle out of the box. Notably, this allows you to execute commands beyond just SELECT statements. A simple way to understand pre-hooks and post-hooks is that they are SQL statements that are executed, respectively, before or after:

- Seeds
- Models
- Snapshots

As with other types of configurations, you can define pre-hooks and post-hooks in both .sql and .yml files. Of course, seeds are an exception to this because they are .csv files, and as such hooks can only be defined in .yml files. In the following, you can see an example of what it looks like to define a pre-hook and post-hook within a .sql file in the config block:

```
{{
  config(
    pre_hook="Some SQL goes here!",
    post_hook="Some SQL goes here!"
  )
}}

select
...
```

Or instead, you could define these hooks within a .yml file. For example, you might want a hook to run for all models in a certain directory, so we could define this in the *dbt_project.yml*:

```
models:
  my_first_dbt_project:
    ...
    marts:
      +pre-hook: "Some SQL goes here!"
      +post-hook: "Some SQL goes here!"
```

The preceding two blocks show you how you can configure a pre-hook or post-hook to run exactly one SQL statement by simply wrapping the statement in quotes to stringify it. However, hooks provide you the flexibility to execute multiple SQL statements or to execute the SQL that is returned from a macro. Throughout this section, we will walk through a couple of practical examples where we use each of these methods for configuring hooks.

Note Notice that the model-level config uses an underscore, "_", as a separator, while the YAML config uses a hyphen, "-", as a separator. This is an error that people new to this syntax commonly run into.

In the following sections, we will walk you through a few practical examples that you can use as reference when implementing pre-hooks and post-hooks in your dbt projects. The first example of this section is a pre-hook that will change database roles, and the second example is a post-hook that is used to apply a Snowflake masking policy. While we show these examples individually, it is important to remember that a model can have both pre-hooks and post-hooks associated with it. So, don't think you are stuck with choosing one or the other.

The examples that we will share with you can be useful in practice, but these just scratch the surface of what you can do with these types of hooks. As with all other examples in this book, we are using Snowflake, and as such some of the operations that we implement will reflect Snowflake-specific syntax. But, the concepts should be transferable across different data platforms such as BigQuery, Databricks, etc.

On a final note, before we start exploring the examples, we caution you to thoroughly think through your use of hooks before implementing them. Hooks are very flexible and allow you to run any SQL statement(s) that you wish. As such, it is easy to fall into antipatterns if you don't carefully think through your use of hooks.

For example, there is nothing stopping you from building a table, or a view, using a pre-hook that runs before a model, but this is certainly an antipattern in dbt because the pre-hook statement could easily be an upstream model on its own. Before using a hook, you should ask yourself if there is another feature of dbt that you could use to implement the code you're trying to run.

Change Database Role with a Pre-hook

For this example, imagine that you are working on a project where the data that you're transforming falls into two categories: sensitive and nonsensitive. The sensitive data that you must process is data for your Human Resources (HR) department, and it includes both Personal Identifiable Information (PII) and Protected Health Information (PHI). The nonsensitive data is everything else in the organization, such as orders, payments, shipments, etc.

Suppose that up until now your dbt project has only ever needed to transform nonsensitive data, but you have a new requirement to process some of the HR data. However, you have found yourself in a situation where the role that you have dbt configured to use doesn't have access to read any of the sensitive data, in this case the HR data. There is a separate database role that does have access to HR data, but that is the only data that this role has access to. Figure 7-1 provides a basic visual of the access rights of these two roles.

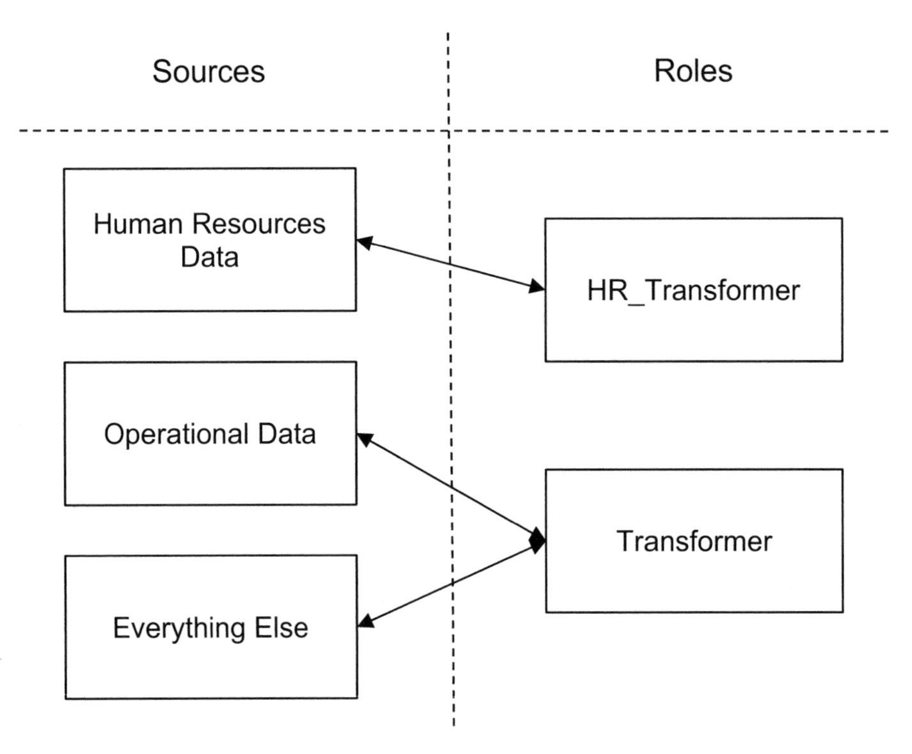

Figure 7-1. *Source data access by role*

There are several approaches that you could take to navigate these requirements, including but not limited to

- Have access to the HR data source granted to the `Transformer` role.
- Create a new dbt project that only transforms HR data.
- Create a new target that uses the `HR_Transformer` role.
- Build the HR data models in the existing dbt project, and swap to the `HR_Transformer` role using pre-hooks before these models start.

For the purposes of our example, let's say that the first three solutions wouldn't work for the requirements of this task. Since we are using Snowflake for our examples, we can easily change database roles, so that leaves us with the fourth option of using a pre-hook to swap database roles before the model starts running.

Caution Swapping roles using hooks has limitations, including the inability to use that role to run tests on the model. For this reason, in a real-world scenario it would be better to use one of the first three solutions listed earlier. With that said, this example will still show you how pre-hooks work and the flexibility that they offer.

The implementation of this solution is rather straightforward, and we will show you two ways that you can configure this pre-hook. The first way would be to implement a pre-hook using model-level configurations. In this example (see Listing 7-1), we have defined a pre-hook, using Snowflake's `use role` command, to switch from the `Transformer` role to the `HR_Transformer` role.

Listing 7-1. Example of using a pre-hook to change database roles for a single model

```
{{
  config(
    pre_hook="use role HR_Transformer;"
  )
}}

select
...
```

While this works for using the HR_Transformer role for the model defined earlier, dbt will default back to the role that you configured your project to use for all other models in your project. If the requirements for processing the sensitive HR data are all satisfied via this one model, then this example will work well, but what if there are multiple models that need to use the HR_Transformer role? There are two options for this:

1. Add this same pre-hook to each model's config block.
2. Move the pre-hook config to the *dbt_project.yml* file.

While option one would work, we don't recommend you do this because you end up repeating yourself N times, where N is the number of models that need to use this pre-hook. This isn't ideal because if in the future you need to modify, or remove, the pre-hook, you will need to update each model where the configuration is being used.

Fortunately, option two solves this problem and allows you to continue following the principles of DRY code. In order to implement option two, you just need to remove the pre-hook from the config block of the individual models and add the pre-hook to the *dbt_project.yml* file as seen in Listing 7-2.

Listing 7-2. Example of using a pre-hook to change database roles for all models in a given directory

```
models:
  my_first_dbt_project:
    ...
    marts:
      ...
      hr:
        +pre-hook: "use role HR_Transformer;"
```

Now, all models within the `hr` directory will have a pre-hook that runs the command to swap to the `HR_Transformer` role.

Mask Sensitive Data with a Post-hook with Multiple SQL Statements

In this example on post-hooks, let's continue with the idea of working with sensitive HR data. Suppose that following the transformation of HR models we need to apply a masking policy to the target tables or views so that certain fields containing sensitive data are not visible to people without the correct permissions.

For our example, we will be using syntax for Snowflake's Dynamic Data Masking feature. We will operate under the assumption that a dynamic masking policy named `hr_sensitive_mask` is already in place, and we will use this policy in conjunction with dbt post-hooks to ensure that it is applied to the correct models. For Snowflake users, the masking policy might look something like Listing 7-3.

Listing 7-3. Example of a Snowflake masking policy

```
create or replace masking policy hr_sensitive_mask as (val string)
returns string ->
  case
    when current_role() in (HR_Analytics) then val
    else '**********'
  end;
```

Note We direct you to your own data platform's documentation to discover the data masking abilities that it may have and for instructions on how to set that up.

Taking a look at the visual in Figure 7-2, you can see the requirements for the `HR_Analytics` and `Analytics` roles should both have access to any HR target tables or views that are generated by dbt. The key difference to note is the way in which the two roles can access the data in these database objects. The `HR_Analytics` role should have full, unmasked, access to the data, while the `Analytics` role should have limited, masked, access to the data.

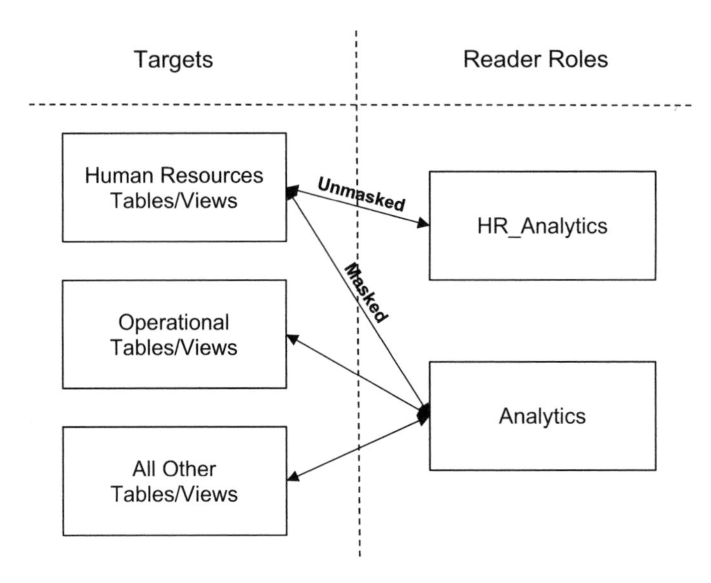

Figure 7-2. *Target database object access by role*

If we apply the masking policy from Listing 7-4. to any dbt models that contain sensitive Human Resources data, we will achieve the access requirements that have been defined in Figure 7-2. To do this, we are going to make use of a set of pre-hooks. This example will serve to show you how you can implement more than one pre-hook at a time.

Suppose that the only table in the HR data mart that contains sensitive data is the model `dim_employees`. There are three fields in the target table that dbt materializes that contain sensitive PII:

- full_name
- social_security_number
- date_of_birth

For the purposes of showing you how you can execute multiple SQL statements using post-hooks, we will apply the masking policy to each of these columns using separate statements. Since there is only sensitive data in one model, we will apply the post-hook configuration at the model level in the dim_employees model. Listing 7-4 shows an example of how we can achieve this by passing a list of SQL statements as an argument to the post-hook configuration.

Listing 7-4. Example showing how to run multiple SQL statements in a post-hook

```
{{
  config(
    post_hook=[
       "alter table if exists dim_employees modify column full_name set
       masking policy hr_sensitive_mask;","alter table if exists dim_
       employees modify column social_security_number set masking policy
       hr_sensitive_mask;","alter table if exists dim_employees modify
       column date_of_birth set masking policy hr_sensitive_mask;"
    ]
  )
}}

select
...
```

As mentioned in the prior chapter, we recommend that you always check for existing packages anytime you are implementing some process that others may have already solved. In this case, there is a package named dbt_snow_mask that can help you set up and maintain masking policies for Snowflake.

On-Run-Start and On-Run-End

In the prior section, we walked through a couple of examples of how you could use pre-hooks and post-hooks, but as you will recall, these types of hooks run before or after an individual model. Having this functionality is useful for conducting administrative activities on tables or views that are materialized from models, but in our experience there are also times that you may need to run some ad hoc SQL when your project starts running and/or finishes running.

Fortunately, the on-run-start and on-run-end hooks can be defined in the *dbt_project.yml* file to be used for exactly scenarios like this. Currently, these hooks run at the beginning and end of the following dbt commands:

- dbt seed
- dbt run
- dbt snapshot
- dbt build
- dbt test
- dbt compile
- dbt docs generate

Earlier in this chapter when we discussed pre- and post-hooks, we covered how to pass SQL directly into hooks, so in this section we will focus on using macros in combination with the on-run-start and on-run-end hooks to run the SQL we need. On-run-start and on-run-end hooks work similarly to pre-hooks and post-hooks in the sense that they can run SQL that is passed in directly as a string, or they can return the results of a macro and run SQL that way if needed. So, just keep in mind that if you need to abstract your SQL to a macro for pre-hooks and post-hooks, you can do that as well.

For this section, we are going to walk through two examples, one on-run-start and one on-run-end, but the two examples will work with each other. The goal is to use an on-run-start hook to generate a table in your target schema that will be used to store the log metadata. Following that, we will create a macro to generate insert statements to write the log metadata to this newly created table and call those macros using an on-run-end hook.

On-Run-Start Example

In this example, we will first build the macro that will be run using the on-run-start hook in the *dbt_project.yml* file. If you recall from Chapter 6, we can define macros anywhere in the *macros* directory of the dbt project by adding a .sql file containing a Jinja macro statement. You can see the macro that we have created in Listing 7-5 or view it at the following path of the book's git repo: `https://github.com/Apress/unlocking-dbt`.

Listing 7-5. A macro that creates a table to log dbt run results

```
{% macro create_logs_table(table_name) %}
    create table if not exists {{ target.database }}.{{ target.schema }}.{{
    table_name }}
    (
        invocation_id string,
        node_unique_id string,
        node_type string,
        result_status string,
        started_at timestamp,
        completed_at timestamp
    )
{% endmacro %}
```

As you can see, this macro will create a table, unless it already exists, with six columns that can be defined as

- **invocation_id**: A unique identifier associated with each invocation of a dbt command
- **node_unique_id**: A unique string for each node in a dbt project, for example, *model.my_first_dbt_project.fct_orders*
- **node_type**: The executed node's associated type, for example, *model, snapshot, seed,* or *test*
- **result_status**: The executed node's associated result, for example, *pass, fail, success, error,* or *skipped*
- **started_at**: The timestamp for when the node execution started
- **completed_at**: The timestamp for when the node execution completed

This is a very simple macro that generates a create table statement, but notice that the macro accepts one argument, `table_name`, which is used to define the name of the log table. We don't need to use the `run_query` function in this macro because dbt will execute the SQL returned from a macro by adding this to the `on-run-start` configuration in the *dbt_project.yml* as seen in Listing 7-6.

Listing 7-6. Example of running a macro via the on-run-start configuration

```
name: 'my_first_dbt_project'
config-version: 2
...
on-run-start: ["{{ create_logs_table('dbt_logs') }}"]
...
```

Now that the macro has been added to the `on-run-start` configuration, anytime dbt is invoked using the commands discussed earlier in this section the DML within the `create_logs_table` macro will be run. Notice that the macro is surrounded by quotations; we do this because behind the scenes the `on-run-start` config expects to receive a string value. As such, by wrapping our macro in quotes, the DML in the macro will be returned as a string and executed by the `on-run-start` config.

Note We used a list to pass the macro to the config, but you can pass just a single string. We only did this so that it is easy to add additional hooks in the future.

With this completed, you could invoke a dbt run to ensure that this is working. For example, if you execute dbt run, you should see a few lines in the console that look similar to those in Listing 7-7. You should find these lines in the console close to the top before any seeds, models, or other nodes are executed and logged to the console.

Listing 7-7. Example console output showing on-run-start info after invoking dbt

```
12:00:00  Running 1 on-run-start hook
12:00:00  1 of 1 START hook: my_first_dbt_project.on-run-start.0 ... [RUN]
12:00:00  1 of 1 OK hook: my_first_dbt_project.on-run-start.0 ... [SUCCESS
1 in 0.59s]
```

Coming away from this example, you should now understand that the `on-run-start` configuration works by running one or more SQL statements at the beginning of the invocation of dbt. Like we did in this example, SQL can be passed to `on-run-start` as a macro. Now that we have the process in place to generate a log table, let's move into an example of using the `on-run-end` configuration to insert records into this table.

On-Run-End Example

Similar to the prior example, we will make use of a macro to generate the SQL for our hook. Though this time, we want this hook to run at the end of each invocation of dbt. As we mentioned at the end of the last section, this is accomplished by using the `on-run-end` configuration. Before we dive into building the macro that will insert the log records into the log table, let's discuss how we can access the metadata that we want to write to our log table.

Recall from the prior section that we create the log table with six columns: `invocation_id`, `node_unique_id`, `node_type`, `result_status`, `started_at`, and `completed_at`. The first field, `invocation_id`, is the easiest to access because it is readily available as a Jinja function *anywhere* within dbt. To get the current invocation identifier, you simply need to call the function like this: `'{{ invocation_id }}'`. The remaining three fields are accessed using a special variable, named `results`, that is only available within `on-run-end` hooks. From this object, you can access the metadata related to each node that was part of an invocation of dbt, including

- Compile and execution timing
- Node unique ids
- Console log messages
- Execution results

This is by no means a fully comprehensive list of what information is available in the `results` object, but the object does contain the data that we need for our log table. If you are curious about what other info is accessible from the `results` object, we recommend you take a look at the *run_results.json* artifact that is found in the *target* directory or print the `results` object to the console using the `{{ log() }}` Jinja function.

Now that you know where you can access the metadata for the log table, let's start building the macro that will insert records into it. The macro can be seen in Listing 7-8 or the following path of the book's git repo: `https://github.com/Apress/unlocking-dbt`.

Listing 7-8. Macro that inserts records into the log table defined with the create_logs_table macro

```
{% macro insert_log_records(table_name) %}
    {% do log(results, info=True) %}
    {% for result in results %}
```

```
        {% set query %}
        insert into {{ target.database }}.{{ target.schema }}.{{
        table_name }}
        values(
            '{{ invocation_id }}',
            '{{ result.node.unique_id }}',
            '{{ result.node.resource_type }}',
            '{{ result.status }}',
            {% if result.timing %}
            '{{ result.timing[1].started_at }}',
            '{{ result.timing[1].completed_at }}'
            {% else %}
            null,
            null
            {% endif %}
        )
        {% endset %}
        {% do run_query(query) %}
    {% endfor %}
{% endmacro %}
```

The macro in Listing 7-8 can be broken down into four parts. The first part is completely optional and is mainly included so that you can see what data is available in the `results` object. We can print the entire `results` object to the console using the log function using this line:

```
{% do log(results, info=True) %}
```

But, you could comment this out or remove it entirely so that the entire `results` object isn't printed to the console every time this macro is called. Following this step, we use a Jinja statement block to begin a for loop to iterate over the results object like this:

```
{% for result in results %}
...
{% endfor %}
```

If you have already explored the data structure of the `results` object, then you may have noticed that it returns a list where each value in that list is another object named `RunResult` which corresponds to each node that was part of the current invocation of dbt. Effectively, each iteration through the results object operates on a single model, snapshot, test, etc.

The final two pieces of this macro happen within each iteration over the `results` object. First, an insert statement is generated and assigned to a Jinja variable named `query`. Of course, this variable could be named anything that you prefer, but we stuck with `query` for simplicity and clarity of the variable's purpose. As you can see in Listing 7-9, within this generated insert statement we access the `invocation_id` function as mentioned earlier, but all of the other values to be inserted into the logs table come from the `results` object.

Listing 7-9. The `query` that will is used to insert records into the log table

```
{% set query %}
insert into {{ target.database }}.{{ target.schema }}.{{ table_name }}
values(
    '{{ invocation_id }}',
    '{{ result.node.unique_id }}',
    '{{ result.node.resource_type }}',
    '{{ result.status }}',
    {% if result.timing %}
    '{{ result.timing[1].started_at }}',
    '{{ result.timing[1].completed_at }}'
    {% else %}
    null,
    null
    {% endif %}
)
{% endset %}
```

Each value lines up with their respective location in the `results` object, and while they are all easy to understand, we want to make a point to mention the two values related to the `timing` array. First, we use an if statement to make sure that there is data in the `timing` array because without this check an error will get thrown if the array is empty.

Next, notice that we are accessing the second value, index 1, of the `timing` array because we want to log the start and end time of the *execution* of the node. If instead we were to access the first value, index 0, we would actually be logging the start and end time of the *compilation* of the node. While this is also useful metadata, it isn't what we want to log in this table.

Lastly, the insert statement that is being stored in the `query` variable will be run using the `run_query` function inside of a do block like you can see as follows:

```
{% do run_query(query) %}
```

Now, it's important to understand what is actually happening here because it might not be what you first expect. Based on what we know about hooks, on-run-end hooks included, we know that they will run the SQL that is passed to them directly or returned to them via a macro. But, since we are using the `run_query` function within this macro, we won't actually be returning any SQL for the on-run-end hook to run. Instead, we use the on-run-end hook as a means to access the `insert_log_records` macro at the end of each execution of dbt. This is useful in the real world because it means that you don't always need to return SQL to the hook from your macros, but instead you can actually execute the SQL directly within the macro. This is favorable in scenarios like inserting records to a log table where you need to run a series of similar but slightly different SQL statements.

Now that the macro is built and you understand what it's doing, to implement it you simply need to call the macro within the *dbt_project.yml* file as seen in Listing 7-10. For the sake of keeping this YAML file tidy, we've placed the on-run-end hook directly below the on-run-start hook.

Listing 7-10. Example of calling a macro from an on-run-end hook

```
name: 'my_first_dbt_project'
config-version: 2
...
on-run-start: ["{{ create_logs_table('dbt_logs') }}"]
on-run-end: ["{{ insert_log_records('dbt_logs') }}"]
...
```

Now that the hook is configured, you can run a dbt command to check that it is working. For example, you could run dbt run, and you should see in the terminal that both the on-run-start and on-run-end hooks were executed successfully. The on-run-end hook message will get logged to the console at the end of all models, tests, etc. and should look similar to the on-run-start message seen earlier in Figure 7-7.

Once you've received a success message in the console, you could navigate over to your database and query the resulting log table using a query similar to the one in Listing 7-11, where `dbt_logs` is substituted for the table name that you used.

Listing 7-11. Example of SQL used to query the log table that has been generated

```
select
    *
from dbt_logs
limit 10
;
```

Once you have explored this data some, you may notice that the `started_at` and `completed_at` fields are null anytime that the `result_status = 'error'`. This is completely expected and is the result of using a Jinja if statement to check the `timing` array when generating the insert statements. Between the two examples that we've explored in this section, you should now have a solid understanding of how on-run-start and on-run-end hooks fundamentally work, what the results object is, and when to use it.

Before we move on, we do want to call out that in a real-world use case, you may want to limit these logging to macros to only run in production. Frequently, it is not necessary to have this level of robust logging in a development environment. We can make a small change to the hook definitions from Listing 7-10 so that they only run when the target name is equal to "prod". Of course, if your production target is named something else, then you will need to update this:

```
name: 'my_first_dbt_project'
config-version: 2
...
on-run-start: ["{% if target.name == 'prod' %} {{ create_logs_table(
'dbt_logs') }} {% endif %}"]
on-run-end: ["{% if target.name == 'prod' %} {{ insert_log_records(
'dbt_logs') }} {% endif %}"]
...
```

Supplementary Hook Context

Between the on-run-start, on-run-end, pre-hook, and post-hook examples, you are now equipped with the necessary knowledge to start building your own processes to run at different points of dbt node execution. However, once you start adding hooks to several places in your project, there is some additional context around hooks that you will want to understand. The first is how pre-hooks and post-hooks behave with respect to database transactions, and the second is the order of operations of hooks. Let's first look at transactions and when they should be taken into consideration when implementing hooks.

Transactions

We expect that you have some understanding of database transactions already, but most briefly transactions are a particularly useful database concept that allow you to isolate individual units of work. Within a transaction, an isolated unit of work can be, and often is, made up of a sequence of DML, DDL, or other SQL command types. The key part of a transaction is that every step will be committed or reverted together; in other words, transactions are an all-or-nothing process.

Certain dbt adapters, notably Redshift and PostgreSQL, allow you to control whether pre-hooks and post-hooks execute inside of the same transaction as the model itself or in a separate transaction. Understanding that transactions are all-or-nothing when it comes to committing changes to the database, depending on your use case there could be times when it is useful to be able to control the transaction which pre-hooks and post-hooks run in. By default, the adapters that use transactions execute hooks within the same transaction as the current running model.

Note dbt's documentation makes a special callout that you ***should not*** use the syntax in the next examples for adapters that don't use transactions by default. This includes Snowflake, Databricks/Spark, and BigQuery.

You can configure hooks to run in a separate transaction either in the config block at the top of the model's .sql file or within the *dbt_project.yml* file. If you want to configure transaction behavior for only the hooks in a certain model, then you will need to set the configuration in the model's config block. This can be achieved in two different ways, using a dictionary and using built-in helper macros, but both methods produce the same result. Listings 7-12 and 7-13 show the syntax for each method, respectively.

Listing 7-12. Example of configuring hook transaction behavior from the config block of a model file using a dictionary

```
{{
  config(
    pre_hook={
      "sql": "Put a SQL statement here!",
      "transaction": False
    },
    post_hook={
      "sql": "Put a SQL statement here!",
      "transaction": False
    }
  )
}}
```

Listing 7-13. Example of configuring hook transaction behavior from the config block of a model file using built-in helper macros

```
{{
  config(
    pre_hook=before_begin("Put a SQL statement here!"),
    post_hook=after_commit("Put a SQL statement here!")
  )
}}
```

Of course, you could also configure hook transaction behavior within the *dbt_project.yml* file as seen in Listing 7-14.

Listing 7-14. Example of using a dictionary format to configure hook transaction behavior from within the dbt_project.yml file

```
name: 'my_first_dbt_project'
config-version: 2
...
models:
  +pre-hook:
     sql: "Put a SQL statement here!"
     transaction: false
  +post-hook:
     sql: "Put a SQL statement here!"
     transaction: false
```

Order of Operations

Now that we've covered how to implement different types of hooks, let's take a moment to understand the order that hooks will execute in. First, let's take a look at the order of operations for hooks within your project. Following this example, we will circle back to discuss how importing packages to your project affects the order which hooks run in.

The order of operations for hooks is visualized in Figure 7-3, where the flow starts with the on-run-start hook.

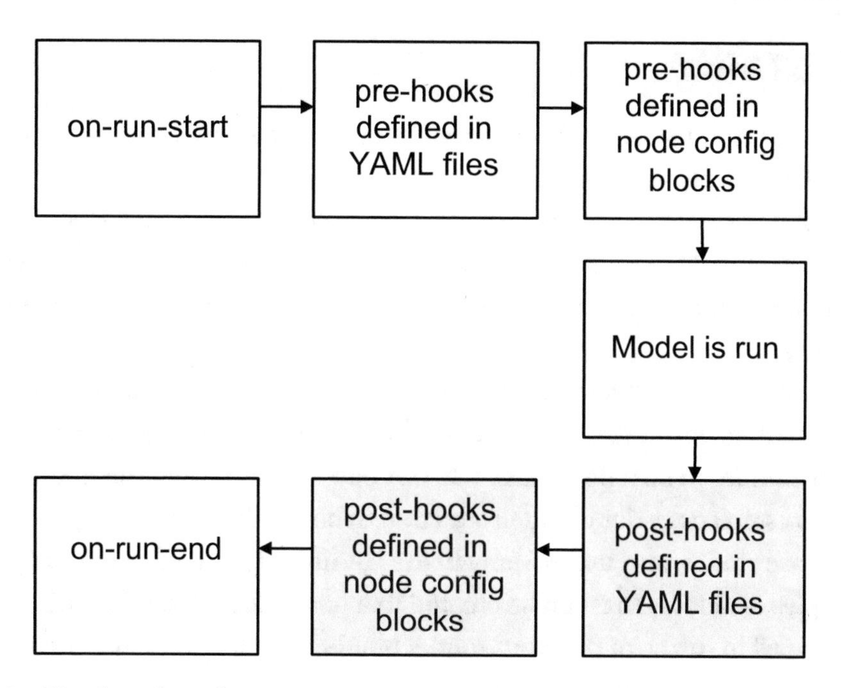

***Figure 7-3.** Hook order of operations*

As you can see, the on-run-start and on-run-end hooks are logically placed at the start and end of the order of operations. With respect to pre-hooks and post-hooks, it might be surprising that hooks defined in YAML files will run before hooks defined within the node config block because the order of priority for model configurations is the opposite; model-level configurations take priority over YAML-defined configurations. However, this actually happens because YAML is parsed before nodes (models, snapshots, etc.), so it does make sense that hooks defined in YAML will run first.

Again referring back to Figure 7-3, notice that the two blocks related to post-hooks have a shaded background. We've done this because post-hooks have unique behavior, and you should be aware that if there is a failure in the node associated with the post-hook, then the post-hook *will not run*. Fortunately, the on-run-end hook will still run even if there are node failures, and this is why when we built our example of logging dbt results to a table, we used the on-run-end hook and not a post-hook. As you can imagine, if we had used a post-hook we would never see records of failure data in the log table.

Warning Post-hooks ***will not run*** following a node failure.

Run Operation

Throughout this chapter, we've discussed different approaches for running different variations of hooks, but each of the types of hook that we've covered runs during some point of a dbt invocation using a command such as `dbt run`, `dbt snapshot`, or `dbt test`. These are all useful within their respective context, but there will be times where you might want to execute some code related to your dbt project without having to actually run any nodes.

This is where the `dbt run-operation` command steps in and enables you to run a macro without having to run any models. This is a powerful component of dbt with lots of opportunities to be able to utilize. In our past, we have used this command to execute macros that copy data from production to our individual developer databases, create database clones, and more. Many in the community have also created a lot of great examples of macros that can be created that have a lot of usefulness, but don't necessarily need to run as part of your model builds. They are more one-off types of solutions you may want to occasionally run.

In our experience, one of the most common use cases for the `run-operation` command is when it's used in combination with the Codegen package. We briefly discussed the Codegen package in Chapter 6 and alluded to the usage of the `run-operation` command, but let's take a moment and use this package as a way to gain an understanding of how to use this command.

In this example, we will use the `generate_model_yaml` macro from the Codegen package to create the beginnings of the YAML for the `fct_orders` table that we created in Chapter 4. To do this, you will first need to make sure that you have added the Codegen package to your project's *packages.yml* file and refreshed your project dependencies by running the `dbt deps` command.

Once you have Codegen installed, you can generate YAML for the fct_orders table by calling the generate_model_yaml macro using the dbt run-operation command. To do this, insert the command from Listing 7-15 into the command line or dbt Cloud IDE, and run it.

Listing 7-15. Example of using the dbt run-operation command to call a macro

```
dbt run-operation generate_model_yaml --args '{"model_names":
["fct_orders"]}'
```

After running this command, you should see results printed to the console that are similar to the following:

```
12:00:00  Running with dbt=1.X.X
12:00:00 version: 2
models:
  - name: fct_orders
    description: ""
    columns:
      - name: orderid
        description: ""
  ...
```

As you can see from the messages that were printed to the console, all that dbt did was execute the contents of the macro that we called and printed the results. It's important to understand that the run-operation command will not execute models, tests, or even on-run hooks.

While using macros from the Codegen package is a great way to get comfortable with using the run-operation command, it doesn't really provide us with any context or understanding of how the run-operation command actually works behind the scenes. Let's work through an example of another common scenario, build a macro to solve the problem, and run that macro using this command.

As we've discussed throughout this book, it is extremely common practice for teams that are using dbt to have one development schema per developer. For example, you may have a team of three people: Andrea, Jeremy, and Martin. Each of these developers will have a development schema in the standard format developer_custom_schema (e.g., andrea_product).

However, each developer may go weeks or even months without working on certain components of the project, so how do we ensure that they are always operating on data that is as close to production as possible? The first, and most simple, solution would be to run the full dbt project anytime the developer feels they need to get their data caught up to production. While for a team of three this might be okay, once your team has expanded to tens or even hundreds of people this will quickly drive your cloud compute costs through the roof!

Fortunately, for Snowflake users we can take advantage of Zero-Copy Cloning to quickly take a snapshot of production tables and clone them into the developer schema(s). The Zero-Copy Cloning feature is a fast and cost-effective way for us to get development environments up to the same state as production. The process clones the metadata of the production tables and shares storage at first, but once you start to operate on the cloned tables, separate storage will be used, so you don't have to worry about overwriting any production data accidentally.

Note We don't cover Snowflake's Zero-Copy Cloning in extensive detail, but we encourage you to refer to their documentation for extended research. For users of other databases, we also encourage you to see if your database provides a feature similar to cloning.

For the purposes of this example, we will assume that the project using this macro has been configured to use a schema naming convention where production schema names only contain the custom schema and do *not* contain the default target prepended to them. For examples of how to achieve this behavior, refer back to Listing 6-9 of Chapter 6.

To implement this, we are going to create a macro name clone_prod_to_dev; this example code can be found in the book's git repository at ~/my_first_dbt_project/macros/dev_env_setup/clone_prod_to_dev.sql, or you can also view this example in Listing 7-16.

Listing 7-16. Macro used to clone production tables to a developer's schema

```
{% macro clone_prod_to_dev(from_database) %}
    {% set clone_queries %}
        select
            'create or replace table ' ||
            '{{ target.database }}' || '.' || '{{ target.schema }}' || '.'
            || table_name
            clone || table_catalog || '.' || table_schema || '.' ||
            table_name || ';' as query
        from {{ from_database }}.information_schema.tables
        where table_type = 'BASE TABLE'
    {% endset %}

    {% set clone_queries = run_query(clone_queries) %}
```

```
    {% for index in range(clone_queries|len) %}
        {% do run_query(clone_queries[index][0]) %}
        {% set current_position = index + 1 %}
        {{ log(current_position ~ ' of ' ~ clone_queries|len ~ ' tables
        cloned to dev.', info=True) }}
    {% endfor %}
{% endmacro %}
```

The easy part here is to actually run this macro, which can be done using the following run-operation command, only you would change the `from_database` argument to be whatever the name of your production database is:

```
dbt run-operation clone_prod_to_dev --args '{"from_database": "analytics"}'
```

While this is easy to run, let's break the macro down into two parts to understand what exactly is happening and why it works so well with the run-operation command. In the first part of this macro, as seen in the following, we create a variable named `clone_queries` that is a dynamic SQL query that is actually querying from the `information_schema` for the production database. Once this variable is set, we use the do function to run the stored query to generate the list of queries that we need to run to clone all of the production tables:

```
{% set clone_queries %}
    select
        'create or replace table ' ||
        '{{ target.database }}' || '.' || '{{ target.schema }}' || '.' ||
        table_name
        clone || table_catalog || '.' || table_schema || '.' || table_name
        || ';' as query
    from {{ from_database }}.information_schema.tables
    where table_type = 'BASE TABLE'
{% endset %}

{% set clone_queries = run_query(clone_queries) %}
```

If you were to substitute the variables in this dynamic query for actual values and run it against Snowflake, you would see a result similar to the results in Table 7-1. As you can see, this dynamic query actually generates a set of queries, generating one query per table in our production database.

Table 7-1. *Example of the queries that are dynamically generated to be run in the clone_prod_to_dev macro*

query
'create or replace table dev.developer_schema.table_x clone analytics.product.table_x;'
'create or replace table dev.developer_schema.table_y clone analytics.finance.table_y;'
'create or replace table dev.developer_schema.table_z clone analytics.product.table_z;'

The second part of this macro runs a for loop of this set of queries that we need to run to clone the individual tables into our developer schema. Within each iteration, we again use the do function to run each query, and then we print the status to the console. We iterate over the length of the `clone_queries` variable so that we can easily print out how far along the process is after each table has been cloned.

While there are a few caveats to this macro (using Snowflake, using singular developer schemas, and views cannot be cloned), the key takeaway here is that when you call the run-operation command you *must* explicitly execute your SQL within your macro. This is necessary because the run-operation command treats returned SQL as a string and nothing else and in turn doesn't run the SQL for you.

Summary

Throughout this chapter, we introduced you to a variety of hooks that can be used throughout your dbt projects. We first discussed that hooks are useful for running ad hoc SQL that doesn't logically fit within another dbt feature such as a model or configuration. We first discussed node-level hooks, known as pre-hooks and post-hooks, which run, respectively, before and after a node has been executed. We used an example of swapping database roles to show how pre-hooks work and an example of applying a masking policy to a table in a post-hook.

While node-level hooks are useful for operating on a singular database object, there are times when it is useful to run some sort of command or query at the beginning or end of a dbt execution. This is where the on-run-start and on-run-end hooks become useful. To better understand how these types of hooks work, we showed two examples that actually worked together to create and write data to a log table so that we can keep record of dbt errors, successes, and other metadata we may be interested in.

With regard to all four of these different types of hooks, we firmly believe that it's important to understand if the design of your hooks is creating an antipattern in your dbt project that you would want to avoid. For example, a common mistake we have seen is for people to use a pre-hook to truncate the destination table before running the model, but this is not necessary because simply using the table materialization would handle this for you. Scenarios like this are something you should try your best to avoid in dbt projects that you build.

To close out the chapter, we discussed the run-operation command which is used to run a macro. This command could be used either in an ad hoc way or be scheduled to run at some interval. Either way, this command provides you a powerful mechanism for running code that is completely unrelated to any dbt nodes. To show how this works, we provided two examples, one using the Codegen package and the second using a custom macro that we built to update our development environment.

Overall, hooks and operations provide you a powerful toolset to run code that doesn't fit well into other components and features of dbt. In the next chapter, we will discuss one of our favorite features of dbt: tests!

CHAPTER 8

Tests

Imagine dedicating weeks or even months to a complex data project, pouring your time and effort into crafting a robust solution, only to witness it crumble shortly after being deployed to production. The frustration mounts as you discover that the data you were so confident about is now riddled with errors and discrepancies. End users, relying on this flawed data, begin reporting issues, shaking the very foundation of trust you had painstakingly established. It's a nightmare scenario that can shatter the relationship between Data Engineers and their consumers.

Traditionally, data quality testing heavily relies on manual checks during the preproduction phase. However, this approach is a recipe for disaster when it comes to analytics. The landscape of applications is in a constant state of flux, always evolving and introducing new bugs that can compromise data quality. The truth is data quality checks are only as reliable as the last time they were executed, leaving room for undetected issues that can wreak havoc downstream. That's why incorporating automated and comprehensive testing into your Data Engineering (or Analytics Engineering) process is crucial.

Enter dbt, a game-changer in the realm of data quality testing. dbt goes beyond its competitors by seamlessly integrating testing capabilities into the model build process. With dbt, you can effortlessly include a wide range of tests to evaluate the integrity of your data transformations. Gone are the days of inadvertently producing flawed or inaccurate data, and no longer will you have to rely on end users to inform you of a broken process.

Throughout this book, we have emphasized how dbt applies Software Engineering best practices to Data Engineering, and testing is no exception. Just as testing is vital in Software Engineering to ensure code functionality, dbt provides the tools to ensure that the data transformations we design yield models that align with business expectations. By leveraging dbt's testing features, you can validate the correctness and accuracy of your data, bolstering trust and empowering data consumers to make informed decisions based on reliable information.

C. Cyr and D. Dorsey, *Unlocking dbt*, https://doi.org/10.1007/978-1-4842-9703-2_8

Why Run Tests?

Before delving into the intricacies of dbt, it is important to understand why testing plays a pivotal role in constructing a robust data warehouse. Regardless of the tool or service being used, comprehensive testing ensures the resilience and accuracy of the code, offering numerous benefits.

First and foremost, **testing instills a deep sense of comfort and confidence in the code you develop**. By subjecting your code to rigorous testing, you gain assurance that it functions as intended and can withstand real-world scenarios. This fosters a positive and assured mindset while working on the project. Testing also serves as a means to validate that your code continues to operate as expected over time. In the dynamic environment of a data warehouse, where data sources and schema may evolve, regular testing becomes imperative to identify and rectify any issues or regressions that may arise. This ensures that your code remains reliable, stable, and adaptable to changes, minimizing disruptions or inaccuracies in downstream data processing and analytics.

Beyond enhancing code reliability, **testing in a data warehouse plays a crucial role in empowering data consumers to make informed decisions based on accurate data**. By subjecting the data and code to systematic tests, you establish a strong foundation of trust and credibility in the platform. This, in turn, enables data consumers to have confidence in the quality of the data they are working with, facilitating meaningful insights and well-informed decisions.

Additionally, **documenting these tests not only serves as a guide for future developers but also encourages collaboration and contributions to the codebase**. By providing clear and comprehensive documentation, you facilitate engagement from both yourself and others in maintaining and enhancing the codebase. This fosters a sense of shared ownership and makes it easier to build upon existing work. Overall, running tests in a data warehouse yields a plethora of benefits, making testing an indispensable practice that drives the overall success and effectiveness of a data warehouse.

When building tests, it is crucial to ensure they are of high quality, regardless of the tool or service being used. So, what constitutes a good test? A good test in a data warehouse possesses several key attributes that contribute to its effectiveness and value. As Ned Batchelder aptly stated, a good test is **automated**, **fast**, **reliable**, **informative**, and **focused**.

Note A good test is automated, fast, reliable, informative, and focused.

Firstly, a good test is **automated**, requiring minimal effort to execute and repeatable in nature. Automating the testing process significantly reduces the time and effort invested in running tests, freeing up resources for other critical tasks. They are also automated as part of the process you are running and not something limited to post-runs.

A good test in a data warehouse is also **fast**, ensuring that the testing phase does not become a bottleneck in the development process. If tests take too long to complete, developers are less likely to prioritize them, compromising the overall quality and reliability of the codebase.

Another essential characteristic of a good test is its **reliability**. A reliable test instills confidence in the accuracy of its results, allowing developers to trust the outcomes and assertions made by the test. Trust in a test's verdict is crucial when determining if something works or does not work, as it enables efficient troubleshooting and debugging. Furthermore, an informative test is invaluable in pinpointing the root cause of errors or unexpected behaviors. By providing meaningful error messages or clues, a test can guide developers toward the specific areas that require attention and correction. This informative aspect expedites the debugging process and enhances the efficiency and effectiveness of code maintenance.

In addition to the aforementioned attributes, a good test in a data warehouse also exhibits **focus**. Each test should be designed to validate one specific assumption or aspect of the code. By focusing on a single validation point, tests become more targeted and precise, facilitating easier identification of issues and swift resolution. This approach also enhances the comprehensibility and maintainability of the test suite, making it easier for developers to understand and modify tests when necessary.

We firmly believe that dbt excels in facilitating all of these attributes within its framework. By leveraging dbt, you can effectively automate, speed up, ensure reliability, provide informative feedback, and maintain focus when testing in a data warehouse, thereby optimizing the overall testing process and contributing to the success of your data projects.

Types of Testing in dbt

In the world of dbt, tests hold a significant role as they serve as assertions about the models and resources within your dbt project. These tests act as gatekeepers, ensuring that your data is not only accurate but also consistent and reliable. When executing the command to run tests (**dbt test**), dbt diligently examines your tests and provides you with clear indications of whether they pass or fail. This crucial step in the development process guarantees the integrity of your data.

Within dbt, you'll encounter two distinct categories of tests: **singular tests** and **generic tests**. Singular tests are SQL queries specifically crafted to identify and return failing rows. These tests are designed to pinpoint any discrepancies or issues that may exist within your data. On the other hand, generic tests take the form of parameterized queries, capable of accepting arguments. This flexibility allows you to create dynamic tests that can be reused across multiple scenarios, increasing efficiency and scalability.

Note dbt allows you to build two types of tests: singular and generic.

In the upcoming sections, we will delve into these two types of tests in greater detail. By exploring their nuances and intricacies, you will gain a comprehensive understanding of how to effectively leverage singular and generic tests within dbt, ensuring the utmost data quality and confidence in your projects.

Singular Tests

Singular tests are the simplest forms of tests to create in dbt. They are just SQL SELECT queries that are written and saved in .sql files within your tests directory that return failing rows. This means that if the query returns any results, then you have failed your test. They are called singular tests, because they are usable for a single purpose.

Let's take a look at a simple example. Suppose that you have a table named "sales" that contains information about each sale made by your company. In this table, you have a column called "price" which represents the price of each sale. You know that every price should be a positive number since you aren't paying customers to take your products. If there is a negative number, then there is a data quality issue that needs to be fixed, and you need to know before downstream users and processes consume the data. In this example, you can create a singular test to check for this. I will start with just writing a SELECT query that lets me know if the price is less than zero. See Listing 8-1.

Listing 8-1. SELECT query that checks if the price is less than zero

```
select
  price
from {{ ref('sales') }}
where price < 0
```

Whenever you run the query, it will either return no results (the test passes) or return results (the test fails). If this query returns results, then we know that we have negative prices in our sales table that need to be corrected. It really is as simple as that.

Singular tests are written very similarly to your models. They are written as SELECT statements and should utilize Jinja commands including the ref() and source() functions in the test definition. The biggest thing that separates these from models is that tests are only stored in the tests directory and they aren't materialized like your models. You can build singular tests for anything you need to test for. The success or failure of the test is just whether or not it returns results.

Because singular tests are so easy to write, you may end up finding yourself using them often. You may even notice that you are writing the same basic structure over and over again for different tests and just changing things like a column name or a model reference. In this case, your singular test is not so singular anymore, and you may want to make it more generic. This is where generic tests come in.

Generic Tests

Generic tests are named appropriately and describe exactly what they are. They are generic tests that can be reused over and over again. Unlike singular tests which are often used for a singular purpose, generic tests are used for multiple purposes. They contain a parameterized query that accepts arguments that provides it with a lot of flexibility.

First, let's take a look at the syntax you need to use to create a generic test:

```
{% test <name of test> (argument 1, argument 2) %}
select
  statement
from {{ argument 1 }}
where {{ argument 2 }} = X
{% endtest %}
```

The {% test %} and {% endtest %} blocks are required so that dbt knows this is a generic test. Within the {% test %} block, you have several fields that you need to input according to your need. First, you need to define the name of your test. This is what will be used later to reference the test, not the name of the actual file like other resources in your dbt project. Additionally, you need to add any arguments (or parameters) that you want to use. Lastly, you need to add your select statement that runs the test. If you define arguments, then be sure they are incorporated into your select statement and are wrapped in curly brackets so that they appropriately conform to the Jinja syntax. In Chapter 6, we reviewed the essentials of Jinja, and everything you learned there is transferable writing tests in dbt.

Now that you understand the syntax, let's take the same example we used for singular tests and look at it in the context of making it generic. Assume we have several other tables in our models that contain numbers that shouldn't be negative. Maybe we have tables containing other values like sales totals, tax, number of shoppers, number of payments used, etc. that we also know should not be negative. We could create singular tests for each one of those similar to what we did for sales, but that is a lot of repeat code. What if, instead, we create one generic test that could execute the same query, but just accept arguments for what is different? In this case, just the column name and model name. We could also set the test name to not_negative. Listing 8-2 shows what the code would look like.

Listing 8-2. Example generic test to make sure the value is not negative

```
{% test not_negative(model, column_name) %}
select
  statement
from {{ model }}
where {{ column_name }} < 0
{% endtest %}
```

In this example, we named our test not_negative. We also created two arguments called model and column_name. When creating generic tests, if you are testing the values in a column then the test should **at minimum** accept these two standard arguments. However, you can pass additional arguments.

Then we created a select query that incorporates our two arguments and looks for values that are negative. Now our example can be used to check for negative values for many model and column combinations instead of having to write this logic multiple times with a singular test. The ability to create arguments is what makes the generic test generic.

This was a really simple example of creating a generic test, but you can get a lot more advanced with it. You can customize your arguments and select statements as much as you like to get the desired results. One interesting note with generic tests is that the model and column_name arguments are special. Whenever you use these arguments, then dbt automatically supplies the values for those when configured in your YAML files. These are the only arguments that do that. Everything else you need to explicitly define.

Note Whenever you create a generic test with model and column_name as arguments, then dbt automatically supplies the values for those when configured in your YAML files.

Unlike singular tests, generic tests need to be added into your sources.yml (for sources) and schema.yml files (for models) to be able to be used. We will cover how to add these later in the chapter.

Out-of-the-Box Tests

dbt comes with some out-of-the-box generic tests that you can use without needing to create anything. These tests are standard items that just about everyone using dbt will need to use at some point. We will look at how we incorporate all the tests we have discussed in the next section.

Within the realm of dbt, you have four predefined generic tests that you can begin using immediately. They are **unique**, **not_null**, **accepted_values**, and **relationships**. These tests come bundled with dbt, providing a solid foundation for validating and verifying the integrity of your data. Let's delve into each of these tests, unraveling their purpose and technical significance in the domain of data testing.

Note dbt comes with four generic tests out of the box: unique, not_null, accepted_values, and relationships.

The first test at our disposal is the "**unique**" test. Its primary objective is to ensure the uniqueness of a column within a model, detecting any instances of duplicate values. By incorporating the "unique" test, data engineers can maintain the integrity and consistency of their data by eliminating redundant entries. This test is often used on the primary key of a model.

Moving forward, we encounter the "**not_null**" test - an essential guardian against the presence of NULL values in a column. This test acts as a robust gatekeeper, ensuring that the specified column does not contain any null values. By employing the "not_null" test, Data Engineers can fortify their data pipelines against unexpected data gaps and ambiguities arising from missing values.

Next on our journey is the "**accepted_values**" test. This powerful test empowers data engineers to enforce a predetermined list of accepted values within a column. By applying the "accepted_values" test, one can guarantee that the data adheres to specific criteria, fostering consistency and adherence to predefined data quality standards.

Finally, we encounter the "**relationships**" test - an invaluable tool for preserving referential integrity. This test diligently examines foreign key values, verifying their alignment with associated references. By incorporating the "relationships" test, Data Engineers can ensure the coherence and accuracy of their data relationships, guarding against inconsistencies and errors in data processing. While this test is useful, we do recommend that you only use it when necessary because it can be a computationally expensive test to run on larger datasets.

These four generic tests - unique, not_null, accepted_values, and relationships - serve as cornerstones of data testing within the dbt framework. They equip data practitioners with the means to bolster data quality, ensuring accurate and reliable data transformations. In the upcoming section, we will explore the seamless integration of these tests, along with other testing methodologies, into your dbt workflow. Prepare to unleash the full potential of data quality assurance and propel your data projects to new heights of reliability and trustworthiness.

Setting Up Tests

In order to let dbt know that you need to run a generic test, you need to configure it in your YAML files. Singular tests will run as part of your model builds, but generic tests need to be told what to run on. They are supported in your models, sources, seeds, and snapshot YAML property files.

Reminder Only generic tests need to be configured in your YAML files. Singular tests will run as part of the model builds.

Let's take a look at Figure 8-1 to see an example model YAML file that contains examples of all of the tests described thus far in this chapter. This is taking all the tests we have discussed thus far and incorporating them into a sample schema.yml file.

```
1    version: 2
2
3    models:
4      - name: sales
5        columns:
6          - name: sales_id
7            tests:
8              - unique
9              - not_null
10         - name: status
11           tests:
12             - accepted_values:
13                 values: ['placed', 'shipped', 'completed', 'returned']
14         - name: price
15           tests:
16             - not_negative
17         - name: customer_id
18           tests:
19             - relationships:
20                 to: ref('customers')
21                 field: customer_id
```

Figure 8-1. *Example model YAML file with tests incorporated*

In this example, we have a model called **sales** and four columns that are part of it: **sales_id**, **status**, **price**, and **customer_id**. For each of these columns, we have different tests configured to show you how they work. Let's walk through each column by column to better understand what is happening.

With the **sales_id** column, we are having dbt run two of the out-of-the-box tests, unique and not_null. This means that dbt is checking the sales_id column (our primary key) to make sure that all values in it are unique and do not contain null values. If any of these assertions are not true, then our project build will fail.

With the **status** column, we are having dbt run an accepted_values check. This is another out-of-the-box test supplied by dbt. This means that dbt is checking to make sure that contents of this column contain the following values: placed, shipped, completed, or returned. If any column has a value that differs from these, then the project build will fail.

Next, we have the **price** column. In this column, we are running the generic test we created earlier in the chapter called not_negative. This custom check is checking to make sure that the value of price is not a negative number. If it does find a negative number, then the project build will fail.

Lastly, we have the customer_id column. In our sales table, customer_id has a foreign key relationship to the customer_id column (or primary key) in the customer table. We know that every value in the customer_id column of this table should have a corresponding customer_id in the customer model, so we want to check that assumption. To do that, we will use the built-in relationship test. For this, we are going to utilize the Jinja ref() function to reference the customer model and then list the field we want to check. If this test finds any customer_ids in the sales table that do not exist in the parent table, then our build will fail.

There is no limit on the number of tests that you can run, so you could add any combination of the preceding tests on any column. We highly recommend taking advantage of utilizing tests in your dbt builds as it is truly one of the main advantages of utilizing dbt.

Configuring Test Severity

By default in dbt, every test will generate a failure when it executes if the conditions of the test are met. Even if there is one failing row, you will still get a failure. More times than not, this is probably what you are looking for. However, other times you may want to run tests against things and have it return a warning instead or make the test conditional on the number of failures returned. Thankfully, dbt makes this possible.

There are three relevant configs that can be used in your YAML files to inform dbt how to handle these. They are severity, error_if, and warn_if. Let's take a look at what each of these is:

- **severity**: This config allows you to select the severity of the test with either error or warn. Error means that the test will fail your build upon error. Warn means that it will add a message to the logs, but the build will continue to run. The default for this setting is error, so if nothing is specified, this is what is used.

- **error_if**: This is a conditional expression that allows you to specify the threshold at which an error will occur. For example, if you are running a duplicate check and only want to get an error if there are more than ten duplicates, then this config can be used to set the threshold. By default, the value here is zero, meaning if there is even one duplicate, you will get an error.
- **warn_if**: This is a conditional expression that allows you to specify the threshold at which a warning will occur. For example, if you are running a duplicate check and you only want to get a warning if there are more than five duplicates, then this config can be used to set the threshold. By default, the value here is zero, meaning if there is even one duplicate, you will get a warning.

These settings can be applied in multiple locations within your dbt project. They can be applied to generic tests, singular tests, and at the project level. Let's take a look at configuring these for each of the levels.

Starting with the generic tests, let's take a look at what this would look like in our schema.yml file. Assume we had a table with a column and we wanted to configure the severity, error_if, and warn_if. We would add a config block under the name of the test along with each of our options. Figure 8-2 shows an example of what this looks like.

```
1   version: 2
2
3   models:
4     - name: my_table
5       columns:
6         - name: my_column
7           tests:
8             - unique:
9                 config:
10                  severity: error
11                  error_if: ">50"
12                  warn_if: ">10"
```

Figure 8-2. *Example model YAML file with test severity incorporated*

> **Tip** A very common place we see severity changed to warn is on tests run in sources.yml file. Oftentimes, you may not be able to change the upstream system to fix the underlying issue, but you do want to be aware it is happening.

With a singular test, you just need to add the config Jinja block to the top of your select statement and define your settings. For example, if you want to fail if there are more than ten returned results, you would add this:

```
{{
  config(
    error_if = '> 10'
  )
}}

select
...
```

Lastly, default settings can be configured at the project level within your dbt_project.yml file. You can specify severity, error_if, and warn_if for various levels of your project, including the entire project, directories, and packages. Again, by default everything is set to error for any results returned for a test. The following example shows how we can change the default severity to be warn for our entire dbt project and how we can set an individual package to error if there are more than ten results returned:

```
tests:
  +severity: warn
  <package_name>:
    +error_if: >10
```

There are a lot of options for how we can configure test severity and lots of layers they can be applied at. We think for most the default behavior is sufficient, especially as you are getting started. But as you get more advanced and want more control over your project, you will likely find instances where you are needing to set severity.

Test Syntax

This section of the chapter is going to focus on the YAML syntax that you need to follow to implement tests. Tests have a supported configuration for models, sources, seeds, and snapshots. We think it is helpful to have this as a reference when you start building assets in dbt. Figure 8-3 shows the syntax for your model YAML files containing tests. With models, tests can be run at the model or column layer.

```
version: 2

models:
  - name: <model_name>
    tests:
      - <test_name>:
          <argument_name>: <argument_value>
          config:
            <test_config>: <config-value>

    columns:
      - name: <column_name>
        tests:
          - <test_name>
          - <test_name>:
              <argument_name>: <argument_value>
              config:
                <test_config>: <config-value>
```

Figure 8-3. *Test syntax for models*

Figure 8-4 shows the syntax for your source YAML files containing tests. With sources, tests can be run at the table or column level.

```
version: 2

sources:
  - name: <source_name>
    tables:
    - name: <table_name>
      tests:
        - [<test_name>](#test_name)
        - [<test_name>](#test_name):
            <argument_name>: <argument_value>
            [config](resource-properties/config):
              [<test_config>](test-configs): <config-value>

      columns:
        - name: <column_name>
          tests:
            - [<test_name>](#test_name)
            - [<test_name>](#test_name):
                <argument_name>: <argument_value>
                [config](resource-properties/config):
                  [<test_config>](test-configs): <config-value>
```

Figure 8-4. *Test syntax for sources*

Figure 8-5 shows the syntax for your seed YAML files containing tests. With seeds, tests can be run at the seed or column level.

```
version: 2

seeds:
  - name: <seed_name>
    tests:
      - [<test_name>](#test_name)
      - [<test_name>](#test_name):
          <argument_name>: <argument_value>
          [config](resource-properties/config):
            [<test_config>](test-configs): <config-value>

    columns:
      - name: <column_name>
        tests:
          - [<test_name>](#test_name)
          - [<test_name>](#test_name):
              <argument_name>: <argument_value>
              [config](resource-properties/config):
                [<test_config>](test-configs): <config-value>
```

Figure 8-5. *Test syntax for seeds*

Figure 8-6 shows the syntax for your snapshot YAML files containing tests. With snapshots, tests can be run at the snapshot or column level.

```
version: 2

snapshots:
  - name: <snapshot_name>
    tests:
      - <test_name>
      - <test_name>:
          <argument_name>: <argument_value>
          config:
            <test_config>: <config-value>

    columns:
      - name: <column_name>
        tests:
          - <test_name>
          - <test_name>:
              <argument_name>: <argument_value>
              config:
                <test_config>: <config-value>
```

Figure 8-6. *Test syntax for snapshots*

All of the available options for testing have very similar options with just some slight differences. Models, sources, seeds, and snapshots all support tests, but at varying levels. As you get started, it is important to understand what your options are. Additionally, it is important to understand the indenting and spacing for the syntax.

Executing Tests

Tests operate similarly to other resources in your dbt project; however, the selection criteria you can use are a little different. This results in you being able to run tests on a specific model, run tests on all models located in a specific directory, or run tests on models that are either upstream or downstream of a particular model. For a full list of the different ways that you can implement node selection, refer back to Chapter 4.

First, let's start with the basics. dbt has a specific command that you run to execute tests called **dbt test**. This is the foundational dbt command to execute tests, and when you run it, all tests in your project are run. If you want to be more selective about what tests are run, then there are lots of additions that you can add to this to specify exactly what you want to do that we will cover later in this section. Before that, we do want to mention that when you utilize dbt build, all tests are run as well.

Note Tests will also run as part of your "dbt build" runs.

Similar to other resource types, tests can be chosen directly by utilizing methods and operators that capture one of their attributes, such as their name, properties, tags, and more. In contrast to other resource types, tests can also be indirectly selected. If a selection method or operator includes the parent(s) of a test, the test itself will also be chosen. This is unique to running the dbt test command and is known as **eager** selection. We will discuss this more later.

First, let's look at the node selection syntax that is available to run with the dbt test. It is actually the exact same as other dbt commands, such as dbt run, and includes --select, --exclude, --selector, and --defer. We covered these in detail in Chapter 4, but let's review what each does:

- **--select**: Used to select a model, or subset of models, to run tests against
- **--exclude**: Used to exclude a model, or subset of models, to run tests against
- **--selector**: Used to run tests using a YAML selector
- **--defer**: Makes it possible to run a test in comparison to a previous state (i.e., run) of dbt

There are also four modes that we want to cover that directly influence the behavior of dbt when running tests. These include how dbt interprets the node selections just highlighted.

- **Direct**: Within dbt, you can directly select which tests you want to run based on attributes such as name, tags, and property types.
- **Indirect – eager**: By default in dbt when selecting any parent resource, a test is triggered as part of an "eager" indirect selection approach. This means that when any parent is chosen, all associated tests will be executed even if they have dependencies on other models. During this mode, if a test relies on resources that have not been built yet, an error will be raised, indicating the presence of unbuilt resources.

- **Indirect – cautious**: This allows users to control whether tests are executed based on the selection status of their parent resources. With this, tests will only run if all their parent resources are selected and built. In other words, if any of the test's parents are unselected, the tests will not be executed. This approach ensures that tests are limited to those whose references fall within the selected nodes, preventing the execution of tests with unselected parent resources.
- **Indirect – buildable**: This is similar to cautious mode, but is slightly more inclusive. This mode specifically includes tests whose references are limited to the selected nodes or their direct ancestors. This broader inclusion of tests becomes valuable in scenarios where confirmation of aggregations having the same totals as their input is required, which involves a test depending on a model and its direct ancestor.

Let's look at some examples of how you can run tests and the type of modes they operate as:

- Run tests on a specific model (indirect):

  ```
  dbt test --select customers
  ```

- Run tests on all models within the "models/marts/sales" directory (indirect):

  ```
  dbt test --select mart.sales
  ```

- Run tests downstream of a model (direct). The "+" after the model name is used to let dbt know to run the downstream models:

  ```
  dbt test --select stg_customers+
  ```

- Run tests upstream of a model (indirect). The "+" before the model name is used to let dbt know to run the upstream models:

  ```
  dbt test --select +stg_customers
  ```

- Run tests on all models with the “Sales” tag (direct and indirect):

  ```
  dbt test --select tag:Sales
  ```

- Run tests on all models with a table materialization (indirect):

  ```
  dbt test --select config.materialized:table
  ```

Viewing Test Failures

Whenever we run our tests, we expect that everything will pass, but what happens when a test fails? How do you know what failed and what to fix? Well without doing anything at all, you can look into the debug logs in dbt and see exactly what failed. You would find the SQL that ran and the error produced that you can use to troubleshoot and figure out your problem. If you are running your build as part of a dbt Cloud job and it fails, then you would still need to look at the dbt logs to figure out what happened. Figure 8-7 shows an example of what failure in dbt Cloud looks like.

Figure 8-7. *Example of a test failure in dbt Cloud*

But there is another way you can also do this. You can have dbt store test failures in the database using the --**store-failures** flag. This will create a new schema in your database and one table per test that stores the rows of data that is causing your failure. It does not store all test failures in the same table since the outputs will be different based on the test. These tables also do not store historical values and are truncated and loaded each time you run them. They are used more to help you troubleshoot an issue, not determine failures over time. You will need to utilize a different method for that, such as the solutions we discussed in Chapter 7.

Note To view test failures, you need to either look at the dbt debug logs or have the results written to a table.

Let's take a look at how you can utilize the store failures feature. The first example is what a command would look like using this if the command was run via the command line:

```
dbt test --select my_model_name --store_failures
```

You can also set this at the individual test level using config blocks. You just set the store_failures value to true or false. By default, this is set to false, so you do not need to explicitly set that unless you have the default set to true in your dbt_project.yml file:

```
{{ config(store_failures = true) }}
```

You can also set this value in your model YAML files for specific tests instead of using the config blocks. For example:

```
1   version: 2
2
3   models:
4     - name: my_model
5       columns:
6         - name: my_column
7           tests:
8             - unique:
9                 config:
10                  store_failures: true  # always store failures
```

And lastly, you can set this in your dbt_project.yml file if you want to set a default for your entire project including models and packages.

```
tests:
  +store_failures: true  # all tests

  <package_name>:
    +store_failures: false # tests in <package_name>
```

You can enable this all the way up to the project level and all the way down to the individual model layer and everything in between.

Test Packages

As we've mentioned throughout this chapter, there are only four generic tests out of the box with dbt. Even though they are all incredibly useful, this is just the tip of the iceberg of what is possible. You can go to town building all the tests you want; however, before doing that, there are some open source packages we highly recommend checking out first. **dbt-utils** and **dbt-expectations** are two very popular and useful ones.

In our opinion, every user of dbt should be using the dbt-utils package because of its immense usefulness. It's a package created and maintained by dbt Labs, so you know it is legit. There are a lot of reasons you should be using this (several shared throughout the book), but for now we are going to focus on just the test portion. As of this writing, there are 16 generic tests that are part of this project. Let's take a look at what these are:

- **equal_rowcount**: Checks that two models have the same number of rows
- **fewer_rows_than**: Checks that the respective model has less rows than the model being compared to
- **equality**: Checks the equality between two models
- **expression_is_true**: Checks to make sure the result of a SQL expression is true
- **recency**: Checks that the timestamp column in the reference model should have data that is not older than the specified date range
- **at_least_one**: Checks that a column has at least one value
- **not_constant**: Checks that a column does not have the same value in all rows
- **not_empty_string**: Checks that a column does not have an empty string
- **cardinality_equality**: Checks that the number of values in one column should be exactly the same as the number of values in a different column from another model
- **not_null_proportion**: Checks that the proportion of non-null values present in a column is between a specified range

- **not_accepted_values**: Checks that no rows match the provided values
- **relationships_where**: Checks the referential integrity between two relations, but with the added value of filtering
- **mutually_exclusive_ranges**: Checks that the ranges defined for a row do not overlap with the ranges of another row
- **sequential_values**: Checks that a column contains sequential values
- **unique_combination_of_columns**: Checks that the combination of columns is unique
- **accepted_range**: Checks that a column's values fall within a specific range

There are some very powerful tests that are part of this that are easily added to your project to use. No need to reinvent the wheel to produce the same tests when they are readily available. This package not only includes several additional tests, but it also enables grouping in tests which adds another layer of granularity. Some data validations require grouping to express specific conditions, as well as others yield more accurate results when performed on a per-group basis. It is just an added feature in how you can execute tests.

The other package that you will definitely want to check out is dbt-expectations. This package is inspired by the Great Expectations package for Python. It was created and maintained by Calogica, which is a data and analytics consulting firm based out of California, and is one we think is extremely valuable when it comes to tests. They have a lot more available tests that we will list out.

Table shape

- expect_column_to_exist
- expect_row_values_to_have_recent_data
- expect_grouped_row_values_to_have_recent_data
- expect_table_aggregation_to_equal_other_table
- expect_table_column_count_to_be_between
- expect_table_column_count_to_equal_other_table

- expect_table_column_count_to_equal
- expect_table_columns_to_not_contain_set
- expect_table_columns_to_contain_set
- expect_table_columns_to_match_ordered_list
- expect_table_columns_to_match_set
- expect_table_row_count_to_be_between
- expect_table_row_count_to_equal_other_table
- expect_table_row_count_to_equal_other_table_times_factor
- expect_table_row_count_to_equal

Missing values, unique values, and types

- expect_column_values_to_be_null
- expect_column_values_to_not_be_null
- expect_column_values_to_be_unique
- expect_column_values_to_be_of_type
- expect_column_values_to_be_in_type_list
- expect_column_values_to_have_consistent_casing

Sets and ranges

- expect_column_values_to_be_in_set
- expect_column_values_to_not_be_in_set
- expect_column_values_to_be_between
- expect_column_values_to_be_decreasing
- expect_column_values_to_be_increasing

String matching

- expect_column_value_lengths_to_be_between
- expect_column_value_lengths_to_equal
- expect_column_values_to_match_like_pattern

- expect_column_values_to_match_like_pattern_list
- expect_column_values_to_match_regex
- expect_column_values_to_match_regex_list
- expect_column_values_to_not_match_like_pattern
- expect_column_values_to_not_match_like_pattern_list
- expect_column_values_to_not_match_regex
- expect_column_values_to_not_match_regex_list

Aggregate functions

- expect_column_distinct_count_to_be_greater_than
- expect_column_distinct_count_to_be_less_than
- expect_column_distinct_count_to_equal_other_table
- expect_column_distinct_count_to_equal
- expect_column_distinct_values_to_be_in_set
- expect_column_distinct_values_to_contain_set
- expect_column_distinct_values_to_equal_set
- expect_column_max_to_be_between
- expect_column_mean_to_be_between
- expect_column_median_to_be_between
- expect_column_min_to_be_between
- expect_column_most_common_value_to_be_in_set
- expect_column_proportion_of_unique_values_to_be_between
- expect_column_quantile_values_to_be_between
- expect_column_stdev_to_be_between
- expect_column_sum_to_be_between
- expect_column_unique_value_count_to_be_between

Multicolumn

- expect_column_pair_values_A_to_be_greater_than_B
- expect_column_pair_values_to_be_equal
- expect_column_pair_values_to_be_in_set
- expect_compound_columns_to_be_unique
- expect_multicolumn_sum_to_equal
- expect_select_column_values_to_be_unique_within_record

Distributional functions

- expect_column_values_to_be_within_n_moving_stdevs
- expect_column_values_to_be_within_n_stdevs
- expect_row_values_to_have_data_for_every_n_datepart

Between the dbt_utils and dbt_expectations packages, there are a ton of potential opportunities with tests here that accommodate a lot of needs. There are always going to be specific business test requirements that dictate custom tests, but we believe this covers a ton of area. To learn more about either of these packages and to get started using them, check out the dbt package hub (hub.getdbt.com) or explore the dbt Slack community.

Other Packages

Testing is a critical component when it comes to utilizing dbt, so there are lots of open source resources available to help with this. In the previous section, we walked through a couple of our favorites, but there are lots more. While we covered packages in detail in Chapter 6, we wanted to mention some of those again that are worth a look when getting started. Not all of these are created and maintained by dbt Labs; however, all of these have been mentioned at some point as part of the official dbt documentation and ones that we have used.

dbt package hub (hub.getdbt.com)

- **dbt_meta_testing**: Package used to enforce that you have the required tests on models in your project
- **dbt_dataquality**: Access and report on the outputs of dbt source freshness and test. Only works with Snowflake
- **dbt-audit-helper**: Package used to perform data audits on diffs

GitHub packages

- **dbt-coverage**: Python CLI tool which checks your dbt project for missing documentation and test
- **dbt-checkpoint**: Python package which adds automated validations to improve your code via pre-commit checks
- **dbt-project-evaluator**: Package which checks your dbt project to see if it is aligned with dbt Lab's best practices

As with anything you pull directly off the Internet, we still advise that you use caution when using these. Anything managed and maintained by dbt Labs is always going to be fine, but others should be subject to some evaluation before immediate integration into your production processes. We have used these without issue before; that doesn't mean that there never will be. Since many of these are maintained by community members or third-party resources, they may not be quick to keep up to date with dbt product updates or bug fixes either, so be patient and graceful if you encounter issues. Or better yet, fix the issue and submit a pull request for them!

Best Practices

As you get started building tests against your models, we have several practical tips that we believe will help you along your journey. These are things that we have learned along the way that we think will help you with building out your dbt testing framework. These are not hard requirements, and we understand that they may not be applicable to every situation, but we do think they will apply to most.

First, we recommend **starting with generic tests and only use singular tests where you have to**. Generic tests can be written once and applied to multiple models and reduce the need to duplicate code saving time and effort with development and maintenance. Only use singular tests when you need something to run for a singular purpose.

Next, we highly recommend **looking at and utilizing open source test packages before you build your own**. Packages like dbt_utils and dbt_expectations are great places to look first, but there are other open source packages you could take advantage of. There is not a lot of value in spending time building something that may already exist. And if it isn't something specific to your business, it probably already does.

Next, we suggest that you **always build your tests as you build your models and not afterward**. It may seem obvious that these things should be hand in hand, but you may have come from a background where you test post-processing as several systems do. Test-driven development is another software engineering best practice, and dbt provides you the tools to follow this. With dbt, testing is something that you should be thinking about as being part of our model creation. Whenever you update your sources.yml files or your schema.yml files, make it a habit to go include tests about assertions you have made of your data.

Lastly, we recommend that you **test your data sources**. This allows you to make sure that assertions you are making about your raw data are true before you start layering models on top of it. We do recommend that you set the severity to warning for your tests, so it doesn't inhibit your transformations from running, but this is something that you should routinely be looking at and communicating with back to your upstream data source owners. Far too frequently, we see data teams implementing weird logic in their SQL models to attempt to circumvent data quality issues that were produced upstream. If you test your data sources, you can instead be alerted on issues early and talk with the data producers to try to find a real solution to the issue.

Summary

In this chapter, we discussed the importance of testing in building a data warehouse and how dbt excels in providing testing capabilities. We highlighted the risks of relying solely on manual checks and emphasized the need for automated and comprehensive testing in data projects. Testing ensures code robustness, accuracy, and adaptability to evolving data sources and schemas, instilling confidence in the codebase and enabling data consumers to make informed decisions.

We introduced two types of tests in dbt: singular tests and generic tests. Singular tests are simple SQL queries that check for specific conditions and return failing rows, while generic tests are parameterized queries that can be reused for multiple purposes, providing flexibility and efficiency. The attributes of a good test were discussed, including being automated, fast, reliable, informative, and focused.

We also covered out-of-the-box tests provided by dbt, such as unique, not_null, accepted_values, and relationships, which help ensure data accuracy and consistency. We explained how to set up tests in YAML files for models, sources, seeds, and snapshots. We also looked at several additional packages that provide a myriad of options of enhancing your testing framework.

CHAPTER 9

Documentation

Documentation is a crucial aspect of any data project, providing essential context and insights for understanding and utilizing your data effectively. However, it is often not given the level of priority that it deserves. Writing documentation is a time-consuming process and one that often doesn't usually bring immediate evident results, whereas developing other things like a new product feature does. It isn't the flashy, sexy work that leaders like to usually see, so it can often be deprioritized. Usually, everyone agrees that it's important, but there always seems to be something else that is more important. Additionally, things change so fast in tech that many choose not to write it out of fear that it will be out of date within a short period of time.

We have long been proponents of documentation living as close to the tools and services that a developer is working with as possible. Though warranted, it is a tough ask for a developer to stop and leave what they are working on to go write a word doc or article using another application. The best documentation is documentation that gets created while the developer is writing code so that it just becomes a normal part of what they are doing. Like other areas discussed throughout the book, this is an area we feel like dbt really excels in. In this chapter, we will explore the power of dbt's documentation features and learn how to create comprehensive and effective documentation for your data models and schema.

With dbt, you can generate clear and intuitive documentation that not only captures the technical details of your data models but also provides valuable insights for both technical and nontechnical users. It does this by collecting information about the data you are working on and your dbt project and storing it in the form of a JSON file. If you are using dbt Cloud, then it has a really nice feature where that information is served up via an easy-to-use web page.

dbt, with a little help from you, automatically extracts metadata for you which collects information about your data models, including object information such as tables, columns, data types, nullability, and even statistics about your data such as row counts and table sizes. It also allows for customization, enabling you to tailor the

C. Cyr and D. Dorsey, *Unlocking dbt*, https://doi.org/10.1007/978-1-4842-9703-2_9

documentation to your organization's specific needs. You can add rich text descriptions to your models, providing detailed explanations and context effectively building out a data dictionary.

The documentation generated by dbt is not just a static representation of your data models and schema. It offers interactive features that empower users to explore and interact with the documentation effectively. Features like data lineage visualization allow users to trace the flow of data transformations, providing a deeper understanding of the data pipeline. Additionally, the ability to view sample data within the documentation gives users a preview of the actual data present in the tables and columns, aiding in data exploration and analysis. Automatic updates ensure that any changes to your data models or schema are reflected in the documentation, eliminating the need for manual updates and saving valuable time and effort.

In the following sections, we will guide you through the process of setting up the documentation environment, generating documentation with dbt, enhancing it with customizations, and maintaining its integrity. We will also explore collaboration and documentation workflows, advanced topics, and the importance of ongoing documentation maintenance. Let's dive in and unlock the power of dbt's documentation features to create comprehensive and informative documentation for your data projects.

Understanding dbt Documentation

To make the most of dbt's documentation capabilities, it is essential to gain a solid understanding of how it works and what it offers. In this section, we will delve into the key aspects of dbt documentation, equipping you with the knowledge needed to create comprehensive and insightful documentation for your data projects.

While we touched on some of these in the chapter introduction, we wanted to highlight each of the benefits of this and describe what it does:

- Interactive web page
- Metadata extraction
- Customization
- Showcase relationships and data lineage
- Automatic updates

One of the biggest benefits of building documentation through dbt with dbt Cloud is that they **create an interactive web page** for you that you can interact with that is accessible directly from dbt Cloud. No need to leave your developer environment, everything is right there where you are working. And by running a simple dbt command (dbt docs generate), you can have this populated and working in a matter of minutes. While there is a lot of enrichment that can be done, you get valuable information immediately populated after starting your project. Figure 9-1 shows what this page looks like once you populate it. We will cover the individual components of this web page later in the chapter.

Figure 9-1. *The dbt Cloud documentation web page*

dbt documentation provides a holistic view of your data models and schema, giving users valuable insights into the structure and relationships within your data infrastructure. By leveraging dbt's **metadata extraction** capabilities, you can automatically extract crucial information such as table and column names, data types, descriptions, and statistics about your objects. This wealth of information enriches the documentation, enabling users to quickly grasp the purpose and content of your data assets.

In addition to capturing the technical details, dbt documentation allows you to add your own **customizations** to it starting with descriptions. This means you can add rich text descriptions about what your tables, columns, and other objects are effectively creating a data dictionary anyone can reference. Also, go beyond the technical specifications and include high-level explanations of the purpose, assumptions, and transformations applied within each data model. These descriptive elements offer valuable context to users, helping them understand the rationale behind the data structures and their role in the overall data ecosystem. You can incorporate additional metadata, such as business glossary terms or data quality indicators, to provide more comprehensive insights. We will cover more of this throughout the rest of the chapter.

Another crucial aspect of dbt documentation is its ability to **showcase relationships and data lineage**. Understanding how data flows and transforms throughout your data models is essential for comprehending the end-to-end data pipeline. dbt's documentation features provide visual representations of these relationships via DAGs, allowing users to trace the path of data transformations and gain insights into the dependencies and impact of changes within the data infrastructure.

The last benefit is that all of your **documentation is automated** and can be built as part of your transformation pipeline. No need to manually populate or update an external page, dbt handles the heavy lifting for you. You still need to keep some information updated in dbt (like descriptions and customized additions), but many items such as metadata fields are automatically updated.

By leveraging the full potential of dbt documentation, you can create documentation that not only serves as a technical reference but also becomes a valuable resource for data consumers and stakeholders. It bridges the gap between technical complexity and user-friendly understanding, empowering individuals across the organization to make informed decisions based on reliable and well-documented data assets.

Adding Descriptions

It is really difficult to take advantage of the great benefits dbt has to offer without utilizing things like sources and tests. Nearly every project will have at least one source YAML file to contain source information. You will likely also have several schema YAML files to contain model information and generic test configurations. These YAML files are where you add descriptions about your objects that get populated into the documentation web page. If this sounds familiar, it is probably because it is. Throughout this book as we covered building YAML files for sources, models, and tests, we have already covered this, so it is coming full circle.

dbt allows you to add descriptions in your YAML files for nearly everything in your project. This includes the following:

- models
- model columns
- seeds
- seed columns
- sources
- source tables
- source columns
- macros
- macro arguments
- snapshots
- snapshot columns
- analyses
- analyses columns

Descriptions can only be added in your YAML files and cannot be created anywhere else. For example, you cannot add a description about a model into the model file itself. You can use comments, but these do not get populated into description fields.

Adding descriptions in your YAML files is really simple, and if you have followed the book to this point, you are probably already comfortable with it; however, let's go ahead and look at an example. Let's suppose I have a model called Sales that has three columns called product, sales_date, and sales_total, and I want to add descriptions to these that get populated into my documentation site. I would open my schema.yml file that contains information about this model (or create one if it does not exist) and add the values. Figure 9-2 shows what this would look like.

models > schema.yml

```
1   version: 2
2
3   models:
4     - name: Sales
5       description: One row per purchased item
6
7       columns:
8         - name: product
9           description: The name of the product that was purchased
10        - name: sales_date
11          description: Date and time that the product was purchased
12        - name: sales_total
13          description: Total amount for the product including sales tax
```

Figure 9-2. *Example of adding a description to the schema.yml file*

The syntax is simple and we just include description: under the name blocks and add our text description. As a general rule of thumb, anywhere you see - **name** in your YAML file, then you can also add a description underneath it.

This is a real simple example where the model name and columns are descriptive already; however, that is not always the case. Even when they are descriptive, it is very hard for a model and column name to tell the full story, and there may need to be additional context shared. For example, we used sales_total which is simple enough, but the user of that data may not know that the field already includes sales tax. Adding a description here provides context to make sure consumers are using the data the same way and understand what it is.

Descriptions do have some really cool capabilities, but also some things that could trip you up if you're not careful. Let's look at a few things to consider when writing your descriptions:

- Descriptions do not need to be contained to one line and can extend as many lines in your YAML file as you like. However, you cannot issue a return to start a new line or it will break.
- You can add multiline descriptions to your models using YAML block notation. This is helpful when you have a longer description that you want to spread out over multiple lines.
- Markdown is supported in the description.

- Your descriptions need to adhere to YAML semantics. If you use special characters like colons, square brackets, or curly brackets, you will need to quote your description. The YAML parser will get confused otherwise and fail. Utilizing Markdown syntax will almost always require quotes.
- You can utilize doc blocks in a description. We will talk about those in the next section. We actually recommend these for anything that is more complicated than a simple sentence or two.
- You can link to other models in a description.
- You can include images in your descriptions, both from your repo and from the Web. We will also cover this in an upcoming section.

Note Your descriptions need to adhere to YAML semantics. If you use special characters like colons, square brackets, or curly brackets, you will need to quote your description; otherwise, you will get a failure.

Doc Blocks

Within your dbt project, you can add any description you want into your YAML files, but sometimes that content can get long. Start adding long descriptions to your YAML files and you will quickly realize that they can get overloaded with information and can become difficult to navigate. Additionally, it is common to have the same definition for a column that is used in multiple models. It would be useful if you could write the description once and then reuse it throughout your project. This is where doc blocks come in to help.

With doc blocks, you create a Markdown file within your project to contain the additional documentation and then just reference it from your YAML files. Markdown files are created by using the .md extension in any resource path. By default, dbt will search in all resource paths for doc blocks, but we recommend keeping them in the same resource directory that you plan to reference them from. Everything still works the same in terms of the end-result documentation, but your YAML files remain a lot cleaner and easier to navigate. Additionally, you can lean more into DRY (Don't Repeat Yourself) practices because you can write a description once and reuse it. As a general rule, we like to only add short text-only descriptions directly into the YAML files. If we need to include a lot of information or need to utilize Markdown, then we utilize doc blocks.

Tip As a general rule, we like to only add short text-only descriptions directly into the YAML file. If we need to include a lot of information and/or need to utilize Markdown, then we utilize doc blocks.

Each Markdown file you create can have as many doc blocks in it as you like, and we typically recommend using it this way. You can create one Markdown file to contain all of your documentation (i.e., docs.md), or you can create multiple Markdown files and group them to your liking. We recommend starting with one and splitting out as the need arises.

Within your Markdown file, you will create the doc blocks. Doc blocks are started with the *{% docs <your doc name> %}*, followed by your text, and then ended with *{% enddocs %}*. These let dbt know how to interpret them. Let's take a look at Figure 9-3 to see what this looks like if we wanted to create a table for order statuses.

```
models > Docs.md

1   {% docs order_status %}
2
3   Once an order is placed, the following statuses are used to describe where it is in the process.
4
5   | status    | description                                  |
6   | --------- | -------------------------------------------- |
7   | Ordered   | The product has been Ordered                 |
8   | Cancelled | The order has been cancelled                 |
9   | Shipped   | The product has been shipped                 |
10  | Received  | The product has been received by the customer |
11  | Returned  | The product has been returned                |
12
13  {% enddocs %}
```

Figure 9-3. *Example of a doc block*

As you can see in this example, we have a table that contains various order statuses and a description about them. We could have taken this text and added it directly into our YAML file, but as you can imagine, it would get really messy. Here, it is really easy to view and understand.

We can write and format the text using Markdown any way we like as long as it is contained within the doc block. In the first line, we had to name our doc block, and this is how we reference it from the YAML file. This one is called order_status, so we can reference it using the *{{ doc() }}* Jinja command. Let's look at Figure 9-4 to see how it is referenced.

```
models > schema.yml
1   version: 2
2
3   models:
4     - name: Orders
5       description: This table contains information about product orders
6
7       columns:
8         - name: status
9           description: '{{ doc("order_status") }}'
```

***Figure 9-4.** Referencing a doc block in a YAML file*

As you can see, we just insert the *{{ doc() }}* Jinja command with the name of the code block, and it does the same thing as it would if we just put it all there. Except now it is a lot cleaner.

Utilizing the same doc.md file that we used earlier, we can create additional doc blocks and reference them using the same Jinja command. We just need to replace the order_status name with the name of our new doc block.

Another benefit of creating Markdown files in dbt Cloud is that it contains a formatter and a preview pane for your Markdown. Other IDEs may have similar features, but it is nice that it is built directly into dbt Cloud. In Figure 9-3, it may look like we created a perfectly formatted diagram, but in reality we just entered the information line by line and let dbt handle the formatting for us. This allows you to format things nicely and see the output of how it will look before saving. While you can utilize Markdown in your YAML files themselves, you cannot take advantage of the formatter and preview features.

Understanding Markdown

Since we just reviewed doc blocks, we are going to take a slight detour to help those that may be new with Markdown. This is not a book on Markdown, so we will only lightly cover it, but do want to give you some insight into what it is and how it is used. If you are comfortable with Markdown already, then feel free to jump to the next section.

Markdown is a plain text formatting syntax that allows you to write content in a simple and readable way while still being able to add basic formatting elements. It was created by John Gruber in 2004 with the goal of making it easy to write for the Web, without the need for complex HTML tags. Markdown files are saved using the .md extension.

Markdown is widely used and supported across various platforms and applications. You can use Markdown to write documentation, create blog posts, format README files on GitHub, write forum comments, and much more. It provides a convenient way to structure your text and add basic formatting without having to worry about the intricacies of HTML or other markup languages.

To get started using Markdown is simple, and everything you do is written in plain text. Let's take a look at some basic syntax and formatting options in Markdown.

Headers are useful for organizing your content and creating different levels of headings. You can create headers by using hash (#) symbols at the beginning of a line. The number of hash symbols determines the level of the header, with one hash symbol indicating the largest header (H1) and six hash symbols indicating the smallest header (H6). Here's an example:

```
# Heading 1

## Heading 2

### Heading 3
```

To emphasize or highlight certain words or phrases, you can use asterisks (*) or underscores (_). For italic text, use a single asterisk or underscore, like *italic* or _italic_. To make text bold, use double asterisks or underscores, like **bold** or __bold__. Here's an example:

This is *italic* and **bold** text.

Markdown supports both ordered and unordered lists. To create an unordered list, simply use hyphens (-), plus signs (+), or asterisks (*) before each item. For an ordered list, use numbers followed by periods (.) before each item. Here's an example:

```
- Item 1

- Item 2

- Item 3
```

1. Item 1
2. Item 2
3. Item 3

Markdown also supports the ability to create tables as we saw in Figure 9-3 earlier in the chapter. These are referenced by using pipes and hyphens to separate headers and columns.

These are just a handful of some of the most common things you can do with Markdown, but not everything. Other options can include things like adding code blocks, tables, task lists, footnotes, and linking URLs. A source that we really like is the Markdown Guide website and specifically their cheat sheet (`www.markdownguide.org/cheat-sheet/`). This cheat sheet provides a quick lookup for how to do most things with Markdown.

Adding Meta

dbt allows you to create descriptions of your objects that allow you to enter any text you like, but you also have an option to create your own metadata for a resource using the meta field. The metadata information is viewable in the autogenerated documentation as its own field and can be anything you want it to be. Examples could be things like owner, contact information, refresh intervals, or whether a field contains sensitive data. But you aren't limited and can add anything you like. In fact, it is also common to add metadata that other tools in your stack might reference, such as your data orchestration platform.

Meta fields are defined in YAML files similar to descriptions, but can also be defined in config() blocks for some resource file types. The following items support the meta field:

- models
- model columns
- seeds
- seed columns
- sources
- source tables

- source columns
- macros
- macro arguments
- snapshots
- snapshot columns
- Singular tests via the config() block

Let's first look at an example of how we could add meta fields to a new source we created called "ProductDB." For this source, we want to create five metadata fields called department_owner, contact, contact_email, contains_pii, and refresh_interval and assign values to them at the source level. Figure 9-5 shows an example of what this would look like in sources.yml file.

```
1   version: 2
2
3   sources:
4     - name: ProductDB
5       meta:
6         department_owner: Product
7         contact: John Doe
8         contact_email: John.Doe@someemail.com
9         contains_pii: true
10        refresh_interval: 6 hours
```

Figure 9-5. *Adding meta fields to a sources.yml file*

You could drill down further and also include meta fields on the individual columns if you chose to. Additionally, we could have set up the same thing in the schema.yml files as well to add meta fields for models and other resource types. The syntax is the same.

You can also assign meta fields directly within your resource files (i.e., models) using the config() block. This can be utilized to assign a new meta value or override an existing value defined in your YAML files. You can do this using the following config block syntax:

```
{{ config(
    meta={<dict>},
) }}
```

The <dict> just needs to be replaced with the name of the meta value and the value that I want to set it to. Suppose I wanted to add a contact to this model and set it to John Doe, then my syntax would look like this:

```
{{ config(
    meta={"contact": "Jane Doe"},
) }}
```

Whenever you add meta fields, they will be generated as separate fields in your dbt docs web page. We will look at how these are viewed later in the chapter when we look at how to navigate the documentation website.

As you can see, meta fields open up a world of possibility in terms of documenting objects in your dbt project and are very beneficial. However, they can get very messy if you don't have a plan in place for them – very similar to systems that utilize tagging if you have ever used one of those. As a result, we do strongly recommend having a standard for how these are used to avoid creating a messy project. The standard should include what meta fields get added, where, and what they are called. Without one, you could end up with lots of meta fields floating around with missing values or different spelling variations of the same thing.

Utilizing Images

dbt supports utilizing images within your documentation, and images can be referenced one of two ways. The first is that you can reference images that are stored directly in your source control repository. The other is that you can reference images via URL links. Images can also be referenced from directly within your YAML files or from doc blocks.

Let's start by talking about storing images directly within your source control repository. In order to do this, you will need to create a new directory in your dbt project called **assets**. This should be created at the root of your project. Once you create the folder, you will need to update your dbt_projects.yml file to configure the assets path. To do this, add the following entry into your YAML file just below your other resource paths. It should look something like this once you add the new entry:

```
model-paths: ["models"]
analysis-paths: ["analyses"]
test-paths: ["tests"]
seed-paths: ["seeds"]
```

```
macro-paths: ["macros"]
snapshot-paths: ["snapshots"]
asset-paths: ["assets"]
```

We only added one new entry, but I added the other entries to provide a point of reference for where it should go. Once added, commit the changes to your repository so the folder is created.

The assets folder is where you can store images that you can reference now. If you are a dbt Cloud user though, then you may try to upload an image directly using the Cloud IDE, and unfortunately this is not supported at the time of this writing. To upload the image file, you will need to add the image directly into the repo outside of the Cloud IDE.

Note Currently, you cannot upload images using the dbt Cloud IDE.

Once you have your file loaded into your project, then you are able to reference it for your documentation. Let's assume I created a file called company_logo.png and loaded it into the assets folder of my repo, and I wanted to reference that in the description of my sales model. To do that, I would open my schema.yml file and reference using the syntax shown in Figure 9-6.

```
1  version: 2
2
3  models:
4    - name: sales
5      description: "![My Company Logo](assets/company_logo.png)"
```

Figure 9-6. *Referencing an image stored in the assets directory*

The square brackets contain the name of the image in my description (My Company Logo), and the directory specifies where the image is stored. Now when I run dbt docs generate, my company logo image will show up in the description.

If I don't want to upload an image into my dbt project, I can also reference images using a URL. Taking the same example we looked at in Figure 9-6, I can just replace the directory (assets/company_logo.png) with a URL. Let's look at Figure 9-7 to see what this looks like now.

```
1   version: 2
2
3   models:
4     - name: sales
5       description: "![My Company Logo](https://mycompanywebsite.com/company_logo.png)"
```

***Figure 9-7.** Referencing an image via a URL*

Within doc blocks, I can get even more creative with this by combining Markdown, text, and images and just referencing those via my YAML files. You just use the same syntax shown here to reference images.

How to Run

Most of the work needed to build in your documentation in dbt is just inputting descriptions about everything. Actually, generating the documentation is a simple process. You just need to run the **dbt docs generate** command. This command is responsible for generating your project's documentation website by doing the following:

1. Creating the website *index.html* file in the target directory of your dbt project. This directory is the location where dbt places generated artifacts including the documentation files.
2. Compiling the project to *target/manifest.json*. This compiles your entire dbt project including models, tests, macros, and other resources into a single manifest file. This file contains metadata about your project's structure, dependencies, and definitions.
3. Produces the *target/catalog.json*. This produces a catalog of tables and views created by models in your project, such as column names, data types, descriptions, and relationships.

Unlike many of the other dbt commands, dbt docs generate does not accept many arguments. The only argument that it does support is the **--no-compile argument** which can be used to skip recompilation (or step 2 of the process defined earlier). When this is used, only steps 1 and 3 are executed. This is useful if you already have a manifest.json file that you want dbt to use to create your documentation.

You can also utilize the **dbt docs serve** command to generate a local documentation site. This command will serve up your documentation site on localhost using port 8080 and opens using your default browser. You still need to run **dbt docs generate** beforehand since this command relies on the catalog artifact that it produces. The dbt docs serve command does introduce some additional arguments that can be used:

```
dbt docs serve [--profiles-dir PROFILES_DIR]
               [--profile PROFILE] [--target TARGET]
               [--port PORT]
               [--no-browser]
```

As you can see, you can specify the profile directory, the profile, and the port (if you want to use something different) and set it to open a browser window.

Local vs. dbt Cloud Documentation

As highlighted throughout the book, there are two types of environments within dbt Cloud. The first is a development environment which is what developers use for development, and the other is a deployment environment which is where production-grade jobs are run. In dbt Cloud, you only have one development environment, but you can have multiple deployment environments (i.e., QA, UAT, Production, etc.). Whenever you are working directly in the dbt Cloud IDE and execute commands via the command line, it is only executed in the development environment. In order to execute commands in deployment environments, we need to create jobs to do that for us. More on this in Chapter 10.

Whenever we execute the dbt docs generate command in our development environment, it creates a local documentation site that only you can see. To access this within dbt Cloud, you need to click the book icon next to where you change your branch to see it as seen in Figure 9-8. You do not click the Documentation tab at the top. The Documentation tab at the top is shared with everyone, whereas your local copy is only viewable by you.

Figure 9-8. *How to view local docs in dbt Cloud*

This is done so that you have a place to review your documentation site before it goes "live" in production. In order to populate the Documentation tab and share with others, you need to do a couple of additional steps.

Whenever you create a new job in dbt, there is a checkbox to generate docs on run that you can select as seen in Figure 9-9. This will automatically run dbt docs generate whenever that job runs creating your documentation site. You can add an individual command line within the job to run this or check the box that does essentially the same thing with one small difference. Whenever you check the box, your job will still be successful even if that step fails. Whenever you add the command as a line item, it will fail your job, and all subsequent steps will be skipped. We recommend just checking the box most of the time and only adding the command when you need to change the behavior.

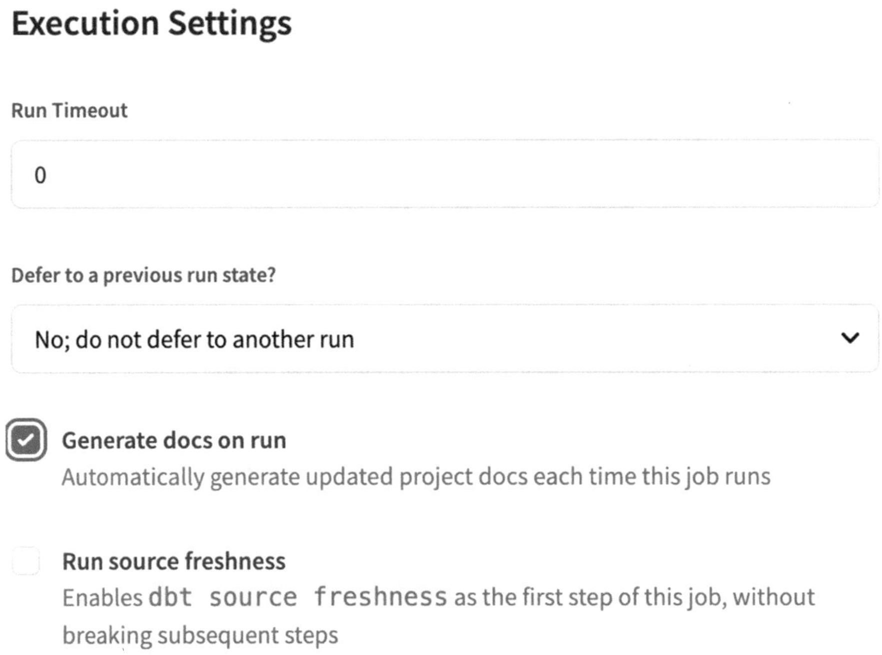

Execution Settings

Run Timeout

0

Defer to a previous run state?

No; do not defer to another run

Generate docs on run

Automatically generate updated project docs each time this job runs

Run source freshness

Enables `dbt source freshness` as the first step of this job, without breaking subsequent steps

Figure 9-9. *dbt Cloud job option to generate docs*

This will run and create a documentation site; however, it will still not generate the shared Documentation tab in your dbt Cloud environment without one more configuration. You need to tell dbt which job will produce the shared documentation. To do this, you need to go into the dbt Cloud account settings and change the documentation artifact. You do this by editing your project details or going to the same place that you edit your warehouse connection and source control repository. There you will find an artifacts section that will allow you to configure the job that populates the shared documentation. See Figure 9-10 for an example of where to find this. Note that this must be a job setup with a deployment environment and cannot be attached to a development environment. It also must be configured to generate docs; otherwise, it will not show up as an option.

Figure 9-10. *Setting the job to configure your shared documentation*

In that example, I have a job called "Prod Full Build" that I am using to populate the shared documentation site. Whenever this job runs now, it will populate the documentation site that everyone can now see. Everything else that runs dbt docs generate will only create a single copy viewable only by the one that is executing the command.

Note Your documentation artifact must be in a deployment environment, and it must be configured to generate docs. Otherwise, it will not show up as an option.

These guidelines apply to deploying documentation when you use dbt Cloud, but if you use dbt Core, the deployment process is a bit more complex because you will need to think through a deployment strategy. We will cover production documentation deployment considerations in Chapter 10, but some common approaches include

- Host a static site on Amazon S3
- Publish with Netlify
- Use your own web server like Apache/Nginx

Value of Reader Accounts in dbt Cloud

Within dbt Cloud, you can have two types of licenses currently, developer and reader accounts. Whenever you purchase licenses, you usually get a certain number of included reader accounts with each developer account that you purchase. But you can also purchase additional reader accounts for a much lower price than the developer accounts if you need more. To be able to view anything in dbt Cloud (including documentation), you must have an account.

The primary reason why you will want reader accounts is so that you can share documentation that is built and hosted in dbt Cloud. You will most likely have folks outside of the development team querying and utilizing your data that needs to make sense of it. They don't need to be developing in dbt, but the documentation produced would be incredibly valuable. In order for them to access it though, they must have an account.

Note The documentation web page that is produced in dbt Cloud is only accessible by users who have a dbt Cloud account.

If you don't want to assign or purchase additional reader licenses for documentation viewers, then there are other alternatives you can look at that we mentioned earlier in the chapter. These essentially involve utilizing other services to populate a custom web page for your documentation to be hosted on.

Navigating the Documentation Web Page

Whenever you generate documentation in dbt Cloud, a documentation web page is generated for you. This section will walk you through what you can do using this web page. In full disclosure, we expect to see this continually enhanced in future dbt Cloud releases, so we recommend checking out the official documentation for the latest news. However, we expect all of the functionality mentioned here to still exist even if the overlay changes.

Whenever you open your documentation page (either the shared or local version) in dbt Cloud, you will see an overview section. The overview section provides information and guidelines for how to navigate the docs site. This is a great place to start if this is your first time using the site. Figure 9-11 shows an example of what this looks like.

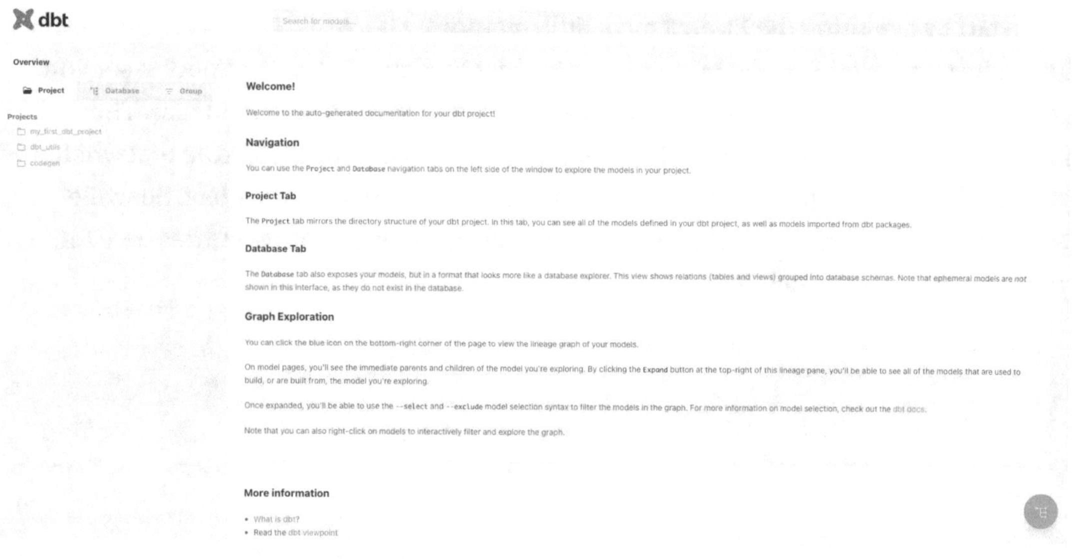

***Figure 9-11.** dbt Cloud docs site overview page*

As you mature your documentation, you may be interested in changing the overview section to something more organization specific, and that is absolutely supported. To change this, you need to create a doc block called "__overview__" to override the existing values. Those are two underscores before and after the word "overview." The doc block can be contained in any Markdown file, but must be named exact to work. Here is an example of what this would look like:

```
{% docs __overview__ %}
<Add your new overview text here>
{% enddocs %}
```

Moving on from the overview page, you will notice that there are two options on the left-hand side of the page called Project and Database. This switch allows you to alternate between exploring your project's folder structure and a database-oriented compilation of tables and views. While the Project view provides convenient accessibility and organization of models, the Database view offers insights into the data presentation for those utilizing the warehouse.

Project Tab

Let's start by exploring the Project tab a little bit more. The Project tab is broken down into subviews called Sources and Projects as seen in Figure 9-12. The Sources section contains an entry for every source you have created in your sources.yml files. The Projects section contains a folder for every project you have created. Note that whenever you install a package to your project, it is interpreted as a separate project. So while you may only be working on one project yourself, you may see other entries here for packages you have installed.

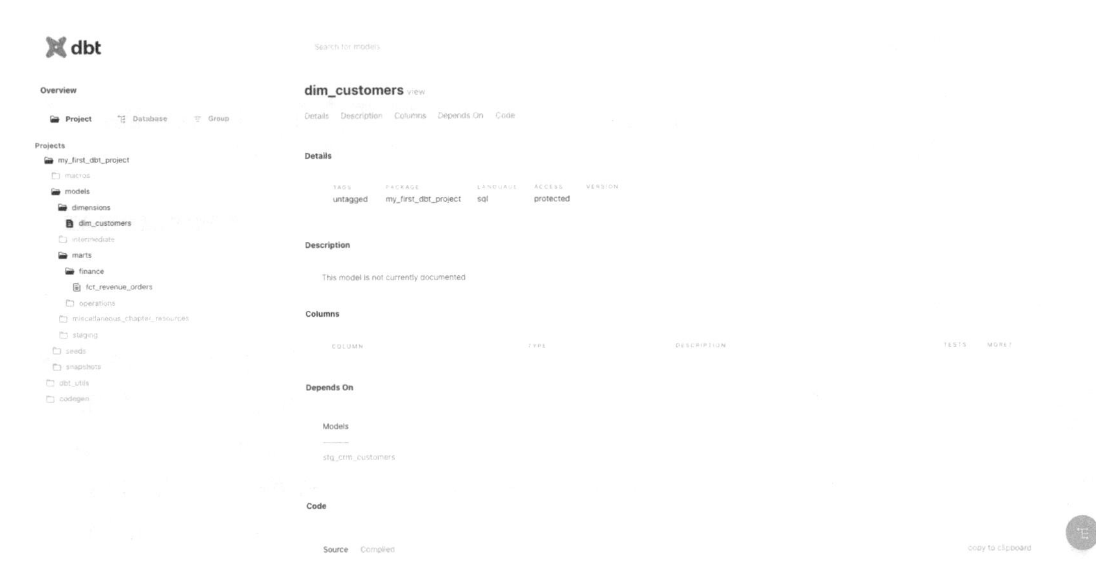

Figure 9-12. *dbt Cloud docs site Projects section*

Whenever you click the Sources, you can view documentation about them. The initial view you see after clicking the source contains the following information:

- **Details**: This includes any meta information you have entered, database and schema names, and number of tables.
- **Description**: Your description of the source if it has one in the sources.yml file.
- **Source tables**: You can see all the tables associated with this source with hyperlinks to view the individual docs about each.

Whenever you click a source in the left-hand pane, you will expand a folder that will allow you to click specific tables to view their documentation. You could also click the hyperlinks from the default source page on the right-hand side of your screen. Whenever you view source tables, you are able to see the following information:

- **Details**: This contains information about the object including tags, owner, object type, and some information about the source. You also get size information, row count, and a last modified date for the object.
- **Description**: Your description of the source table if it has one in the sources.yml files.
- **Meta**: Any meta fields that you have created and their values.
- **Columns**: This contains a list of all the columns and includes column names, data types, descriptions (if entered into the sources.yml files), and a list of all tests that run against that column.
- **Referenced by**: This shows all of the models and tests that this column is referenced in along with hyperlinks to view them.
- **Code**: This includes a sample SELECT command to query the column.

Moving on to the Project subview, you will see a list of all of your projects. You may only see one or may see many depending on the number of projects you have created and packages installed. Each of the project folders is structured similarly to the dbt project you develop in and separated out based on the resource folders (i.e., macros, models, seeds, tests, snapshots, etc.).

Whenever you click a project link, the first view you see is an overview section similar to the one you saw when you first opened the docs site. By default, every project will look the same, but you are able to change these as well. To set custom project-level overviews, you just need to create a doc block named __<Project Name>__. Those are two underscores before and after the project name. The <Project Name> should be replaced with the name of the project. Let's assume you have a dbt project called "Sales" and want to create a custom overview page. You would create a doc block that looks like the following to override the default setup:

```
{% docs __Sales__ %}
<Add your new overview text here>
{% enddocs %}
```

You can do this for every project in your environment or leave it as is. The choice is yours. Whenever you drill down further into the project subview, you will see individual resource folders that you can further drill down into until you get to individual objects. The documentation could change based on the type of resource you are looking for, but you essentially are able to see the following things:

- **Details** (models and seeds only): This contains information about the object including tags, owner, object type, and some information about the source. You also get size information, row count, and a last modified date for the object.
- **Description**: Your description of the object if it has one in your schema.yml files.
- **Meta**: Any meta fields that you have created and their values.
- **Columns** (models only): This contains a list of all the columns and includes column names, data types, descriptions (if entered into the schema.yml files), and a list of all tests that run against that column.
- **Referenced by**: This shows all of the models and tests that this column is referenced in along with hyperlinks to view them.

- **Code**: This includes a copy of the code that is used to produce the object. For example, if this is a model, then it will show you the exact code from the model. You can also view the compiled code or example SQL for certain resource types. The compiled code is where everything is translated to SQL and could be copied and run directly against your warehouse.
- **Arguments** (macros and tests only): Details about the arguments for macros and tests.

Database Tab

The Database tab shows the same information as the Project tab, but in a format that looks more like a database explorer. The explorer looks at objects as databases, tables, and views as you can see in Figure 9-13

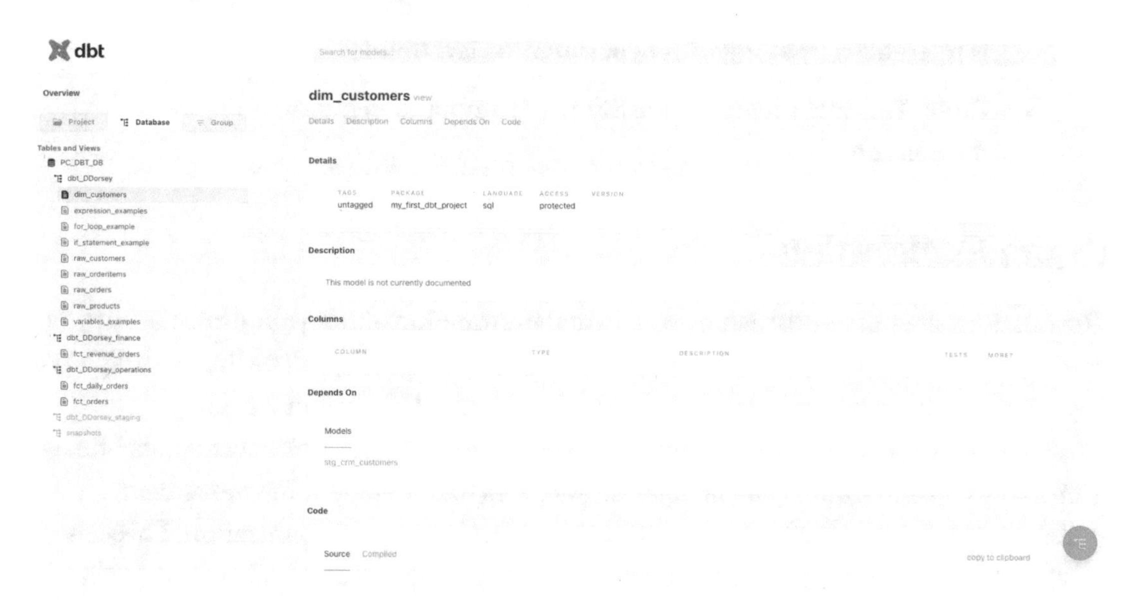

Figure 9-13. *dbt Cloud docs site Database tab*

On this tab, you can only see information on tables and views. You can see the database and schema names, but when you click them, there are no pages that come up. They are only there to allow you to navigate down to your tables and views.

When you drill down to individual tables and views, you have access to the following information about them:

- **Details**: This contains information about the object including tags, owner, object type, and some information about the source. You also get size information, row count, and a last modified date for the object.
- **Description**: Your description of the object if it has one in your sources.yml or schema.yml files.
- **Meta**: Any meta fields that you have created and their values.
- **Columns**: This contains a list of all the columns and includes column names, data types, descriptions (if entered into the schema.yml files), and a list of all tests that run against that column.
- **Referenced by**: This shows all of the models and tests that this object is referenced by along with hyperlinks to view them.
- **Code**: This includes a sample SELECT command to query the column.

Graph Exploration

The dbt docs site not only provides you textual information about your projects, but you can also view relationships within your project via a lineage graph. The lineage graph is a DAG that allows you to view dependencies of your model in a graph format.

Before diving into how to view this in the docs site, let's first look at what data lineage is and why it is important. Data lineage allows users to understand the movement, utilization, transformation, and consumption of data within an organization. This can be visually represented via a DAG (or lineage graph) or via a Data Catalog. In the case of dbt docs, you actually get both. Understanding the data lineage allows engineers to construct, troubleshoot, and analyze workflows with improved efficiency. Additionally, it allows data consumers to comprehend the origins of the data.

Data and how we use data are always changing, so it is simple to understand why we need to understand data lineage. There are three main reasons why understanding this is important:

- Fixing data issues quicker by being able to quickly identify the root causes. With data lineage, you can quickly work backward from the problem to determine where the failure is.
- Understanding downstream impact on upstream changes. Whenever a development team needs to make an upstream schema change (like dropping or renaming a table), you can quickly identify the downstream impact.
- Providing value to the team by creating a holistic view of the data that folks can use to better understand what is happening; also, it helps prompt clean pipelines by easily identifying redundant or messy data flows.

Now that we covered some basics on why data lineage is important, let's switch back to the lineage graph that comes with dbt. To access the lineage graph view, you just need to click the option in the docs site. Currently, this is accessible via a teal box on the bottom-right corner of the docs web page.

Depending on where in the docs site you click to go to the lineage graph will affect your default view. If you are in the docs and looking at a specific model and go to view the graph, it will default to only showing you the model plus the upstream and downstream dependencies. No matter what default view you get though, you can also filter and customize it to show what you are looking for. Figure 9-14 shows a simple view of what this looks like.

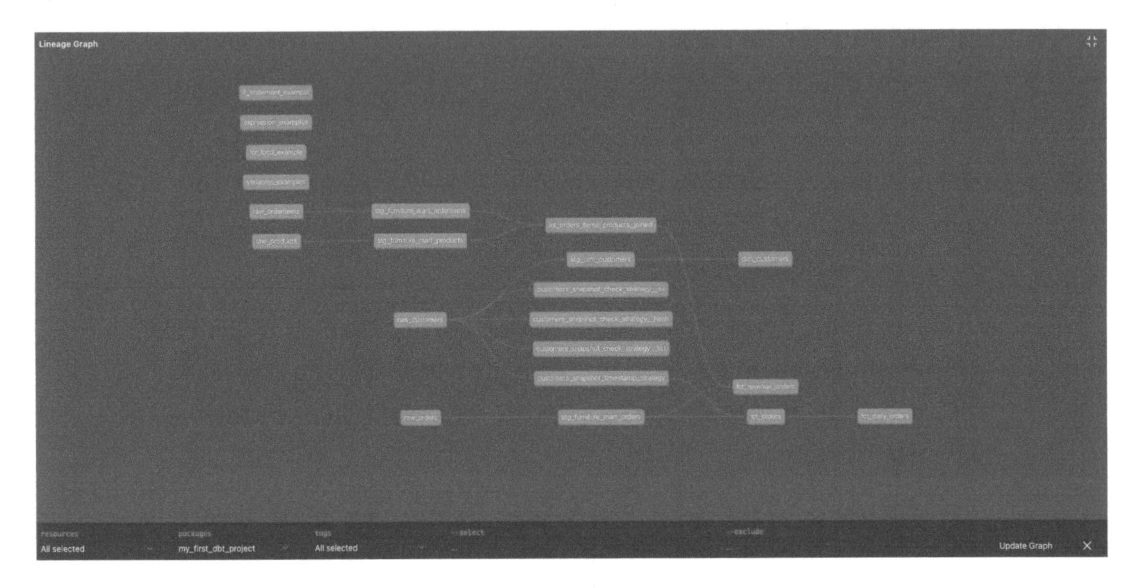

***Figure 9-14.** dbt Cloud docs site lineage graph*

The lineage graph is very interactive, and you can move objects around to create your perfect visual. You can also select any object in the lineage graph, and it will highlight all dependencies. This is really nice when you are trying to visualize the dependencies on a certain object.

You also have several filters that can be applied to narrow or expand the scope of your lineage graph. The following are options that you can use to do this:

- **resources**: Here, you can select what type of resources you want to see on your graph. You can select models, seeds, snapshots, sources, tests, analyses, exposures, metrics, and all.
- **packages**: Here, you can select the packages or projects that you are interested in seeing. You can select individual ones or select all.
- **tags**: If you are utilizing tags in your project, you can select/deselect ones here.
- --**select**: This is a free text field that allows you to type in specific objects you are looking for. This works the same as using --select in a dbt run command.
- **--exclude**: This is a free text field that allows you to exclude specific objects you are looking for. This works the same as using --exclude in a dbt run command.

There are many great third-party tools that have lineage capabilities, but these often come with a hefty price tag. If you aren't already using something else, then we believe the lineage capabilities in dbt are more than enough for most companies, and you should start there before exploring other options. After all, it is included in your dbt Cloud licensing costs.

Searching the Docs Site

The last component of the dbt docs site that we want to cover is the search feature. While you can absolutely navigate through the explorer to find what you are looking for, you can also just search for it using the search bar at the top of the docs site. Figure 9-15 shows an example of the search bar.

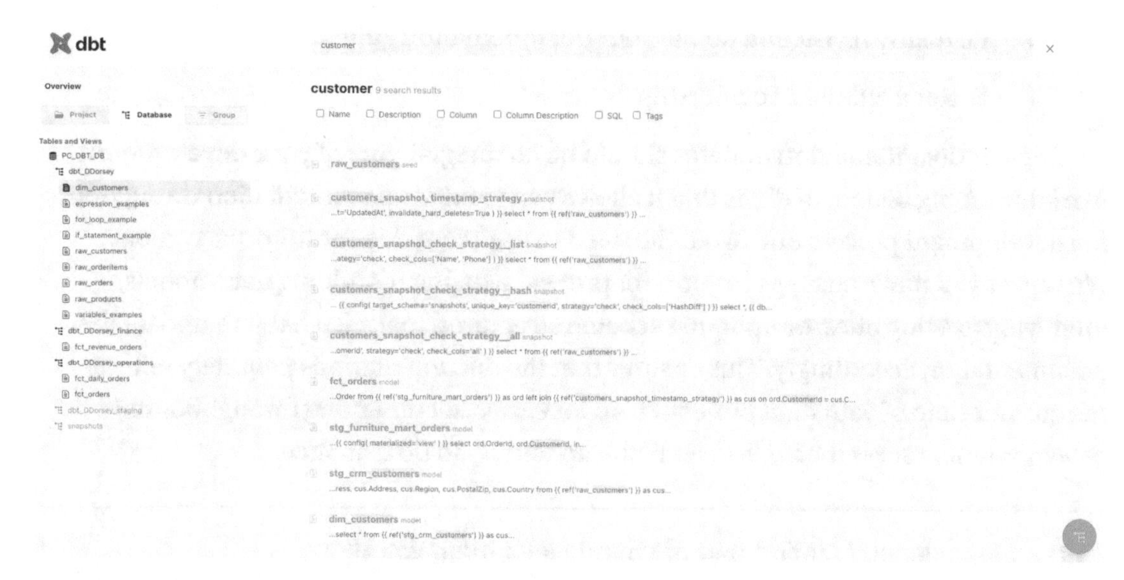

Figure 9-15. *dbt Cloud docs site example search*

You can start typing anything you like into the search bar, and dbt will show anything that contains the words you are searching for. This includes names, descriptions, columns, code, and tags. You can also choose to search based on name, description, column, SQL, and tags if you know you are searching for a specific thing. As your project expands, being able to filter comes in handy to prevent overloaded results.

Maintaining Documentation

Maintaining up-to-date documentation is crucial for ensuring its relevance and usefulness. As your dbt project evolves, it's essential to have processes in place to keep your documentation in sync with the latest changes. This section will explore strategies and best practices for keeping your dbt documentation up to date. At a high level, let's look at some considerations that can improve your ability to maintain your documentation and then dive into each throughout this section:

- Make it part of your development workflow
- Create a documentation review process/make it part of code review
- Encourage open channels of communication
- Regularly revisit and define your documentation guidelines
- Foster a culture of ownership

Regular documentation updates should be an integral **part of your development workflow**. A big benefit of dbt is that it allows you to create documentation throughout the development process and avoid the need to use Word docs or third-party tools. Whenever you make changes to your dbt project, whether it's adding new models, modifying existing ones, or updating schema structures, make it a habit to update the documentation accordingly. This ensures that the documentation accurately reflects the current state of your data project. It is a lot easier documenting throughout the development process than it is to let it pile up and try to do it at once.

Tip Make documentation part of your development workflow.

Consider establishing a **documentation review process** within your team. Before merging code changes or deploying new features, include a step where the documentation is reviewed for accuracy and completeness. This review process helps catch any discrepancies or missing information, ensuring that the documentation remains reliable and up to date. In the next chapter, we will cover the code promotion process with source control which includes mandatory code reviews, and we believe this is a great place to check to make sure new or changed items are properly documented. The code reviewer should be checking it before approving a pull request.

Communication and collaboration are key when it comes to keeping documentation in sync. **Encourage open channels of communication** among team members, especially when making significant changes to the data project. Ensure that any updates or modifications to the project are communicated to relevant stakeholders, keeping everyone informed and aligned. Documentation is great, but it is not a substitute for properly communicating with others about changes.

Regularly review and refine your documentation guidelines and standards. As your data project evolves and matures, you may discover new practices or requirements that need to be captured in the documentation. By periodically revisiting your documentation guidelines, you can adapt them to reflect the evolving needs of your data project and ensure that the documentation remains comprehensive and aligned with best practices.

Finally, **foster a culture of documentation ownership within your team**. Encourage team members to take ownership of the documentation related to their specific areas of expertise. This includes not only maintaining their own documentation but also collaborating with others to ensure cross-functional coverage and accuracy. Documentation ownership instills a sense of responsibility and accountability, leading to more robust and reliable documentation.

By implementing these strategies and best practices, you can establish a documentation workflow that keeps your dbt documentation in sync with your evolving data project. Remember that documentation is an ongoing effort, requiring regular updates and collaboration. With a good documentation process, you can maximize the value of your dbt documentation and empower users to make informed decisions based on reliable and up-to-date information.

Codegen Package

Throughout the book, we have discussed open source packages that you can take advantage of, and one of the best ones for documentation is the Codegen package. The Codegen package can be found on the dbt package hub and one we consider a must-have.

The Codegen package is essentially a YAML templating tool when it comes to adding descriptions about your sources and models. As you can imagine, it can take a lot of time just to input all your tables and columns. Not just in adding in your descriptions themselves but also manually structuring the YAML file to include tables, columns, and descriptions. Take a look at Figure 9-16, and you will see an example of adding one source with a couple of tables and columns. This only took a few minutes to write out, but imagine if I have lots of sources with hundreds of tables and thousands of columns that I also want to do this for. An engineer could spend hours or days just creating syntax just to get to the point they can add descriptions.

```
1   version: 2
2
3   sources:
4     - name: source_name
5       description: ""
6       tables:
7         - name: table_a
8           description: ""
9           columns:
10            - name: column_1
11              description: ""
12            - name: column_2
13              description: ""
14            - name: column_3
15              description: ""
16            - name: column_4
17              description: ""
18            - name: column_5
19              description: ""
20            - name: column_6
21              description: ""
22            - name: column_7
23              description: ""
24            - name: column_8
25              description: ""
26            - name: column_9
27              description: ""
28            - name: column_10
29              description: ""
30        - name: table_b
31          description: ""
32          columns:
33            - name: column_1
34              description: ""
```

Figure 9-16. *Setting the job to configure your shared documentation*

No engineer wants to just do what amounts to data entry. Instead, I can use the generate_source and generate_model_yaml macros in the Codegen package to build all of this out for me. When run, both of these macros generate the templates for you, so you can just focus on adding in descriptions.

This is a package that is created by dbt Labs and one we don't think will ever go away. However, we could see dbt eventually integrating this functionality into their core product eliminating the need.

We have talked about various packages in Chapter 6 and throughout the book, and there are others that help (sometimes indirectly) with generating documentation, and we highly recommend taking advantage of those if any way they can help. As a general rule with dbt, if you are having to remember to do something or do something manual or repetitive, then there is a very high chance something was built to address that need. Always check out the dbt package hub to see what is available.

Tip If you are having to perform a manual and repetitive task in dbt, then there is a very high chance that someone has built something to automate that. Always check out the dbt package hub to see what is available.

Summary

You have now completed your journey through the world of dbt documentation. We have covered the essential aspects of creating, managing, and enhancing documentation for your dbt projects. By following the best practices and techniques outlined in this chapter, you are well equipped to create comprehensive, user-friendly, and valuable documentation.

In this chapter, we learned about the vital role that documentation plays in ensuring the success of your dbt projects by empowering users to understand the data infrastructure, promote collaboration, and facilitate knowledge sharing. Specifically, we looked at the features dbt provides us including automated metadata extraction and customizations such as descriptions, custom metadata, Markdown, images, and more that can be used to populate the dbt Cloud docs site. Then we looked at how to populate our documentation. We also looked at how to navigate and search the docs web page, including the project, database, and graphical views it supplies.

Remember that documentation is not a one-time task but an ongoing process. It requires continuous effort, regular updates, and a commitment to maintaining its quality and relevance. So while dbt provides us with some amazing capability here, it will take some effort from you and your team to make it work.

CHAPTER 10

dbt in Production

As you've learned throughout this book, dbt is a fantastic tool for building, maintaining, and scaling your data transformation processes. We've covered many aspects of dbt ranging from utilizing seed files, building incremental models, and using Jinja and macros to fill in the gaps where SQL gets repetitive. So far, we have only run these transformations in a development environment, but once you are ready for data consumers to utilize the data models that you have built, you will need to deploy dbt to a stable production environment.

It goes without saying that your development environment should not be responsible for handling production workloads for any software product, and data products produced by dbt are no different. There should never be a single point of failure when it comes to production dbt runs, but having data consumers use data being produced by jobs running in your development environment violates this rule. Throughout this chapter, we will discuss how to design deployment strategy to ensure that transformed and validated data is available for use by various stakeholders, such as analysts, data scientists, and anyone else who might need to access the data.

There are different aspects to running dbt in production, including

- Understanding what environments are and how they differ between open source and hosted versions of dbt (dbt Cloud).
- Being aware of the trade-offs of using dbt Cloud vs. dbt Core for a production deployment. This topic will come up many times throughout the chapter.
- Scheduling dbt production jobs.
- Implementing CI/CD pipelines to automate the production deployment process.

C. Cyr and D. Dorsey, *Unlocking dbt*, https://doi.org/10.1007/978-1-4842-9703-2_10

We will cover each of these topics in this final chapter, and by the end you should feel comfortable with having the basic knowledge needed to deploy dbt to a production environment.

Understanding Environments

Before we dig into how environments work within the context of dbt, more specifically the differences in environment behavior in configuration between dbt Cloud and dbt Core, let's briefly cover what environments are. An environment refers to a specific configuration or setup where dbt is executed, deployed, or tested. An environment represents a distinct context that replicates the conditions necessary for developing, testing, and running dbt. Environments are typically created to support different stages of the software development lifecycle, including development, testing, staging, and production. Each environment serves a specific purpose and may have unique characteristics, configurations, and constraints.

Git Workflows and dbt Environments

As a best practice, the number of dbt environments you have should correlate directly with your git workflow strategy. There are two git workflow strategies that we see teams be very successful with: Gitflow and Feature Branch Workflows. There are some very strong opinions from members of the industry about which git workflow is the best to follow, but we will stray away from our subjective opinion and instead focus on how you can implement either of the two git workflow strategies that we've mentioned, starting with feature branching.

Feature branching uses one shared branch, often named main, where developers create their feature branches directly off of it. Developers work directly in their feature branches, and when they have completed their work, they will merge their branch into the main branch. This is often the most simple git workflow that a team can utilize in conjunction with dbt because it is easy to understand and to maintain. Figure 10-1 shows an example of a git tree for a team using feature branching.

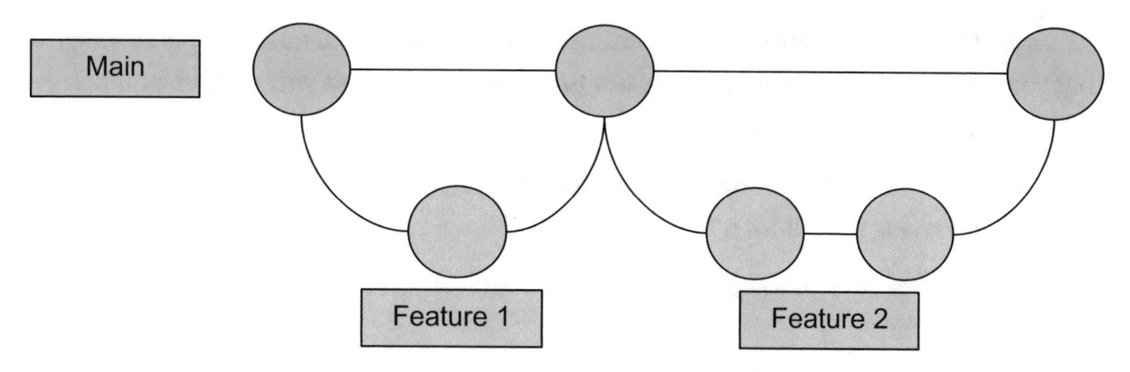

Figure 10-1. *Example of feature branching*

When using a feature branching strategy, you will only need to maintain two environments: a production deployment environment and a development environment. The production environment will correspond with the main branch. When production dbt jobs run, they should be using the code from the main branch because it represents the most recent and up-to-date code. Secondly, you will need to maintain a development environment. The development environment is where developers can work on their new features without having to worry about interrupting the production system. Most often, the development environment points to a different target database than the production environment, but we will discuss this more shortly. Figure 10-2 provides an expansion of the previous image to show which branches belong in which environment.

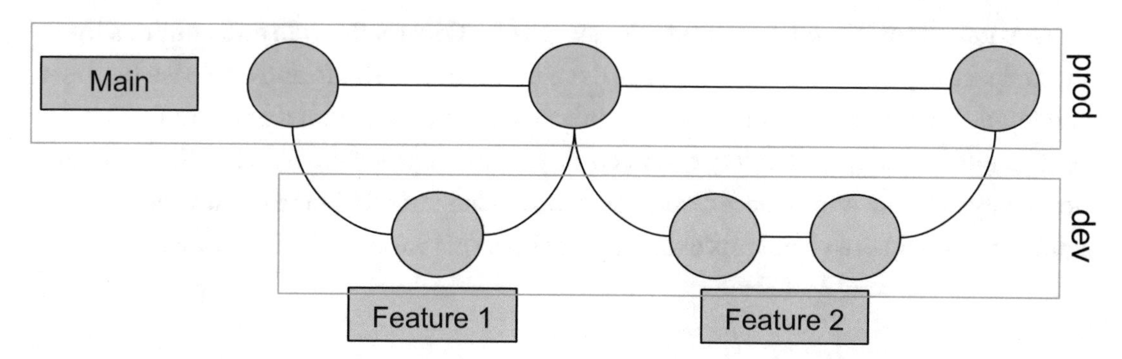

Figure 10-2. *Example of dbt environments when feature branching*

As we mentioned before, this branching strategy is one of the most simple strategies that you can implement, and for many dbt projects, this can be a successful workflow. This strategy works best when you

- Want to deploy changes to production when they are made, instead of having to wait for a release cycle to end
- Need to be able to iterate quickly without having to manage extra git branches
- Are interested in setting up fully automated deployments
- Don't need your changes to go through rigorous user testing

The second type of branching strategy that we will cover is Gitflow. In the context of dbt, this branching strategy will be a multideployment environment strategy. The basics of Gitflow include having two core branches: develop and main. The develop branch serves as the integration branch for ongoing development work. It contains the latest development code and acts as the base for creating feature branches. When work in feature branches has been completed, the code is merged into the develop branch.

Note The develop branch can be named QA, UAT, or anything else, but develop is the most common.

The main branch contains the stable and production-ready code and contains the latest release of your dbt code. When the develop branch is stable and a release has been scheduled, code from this branch will be merged into main so that a new production deployment can happen. Gitflow can become more complex than this, but the additional complexity of hotfix and release branches typically isn't needed for a dbt project. Figure 10-3 provides a visual representation of how this variation of Gitflow works.

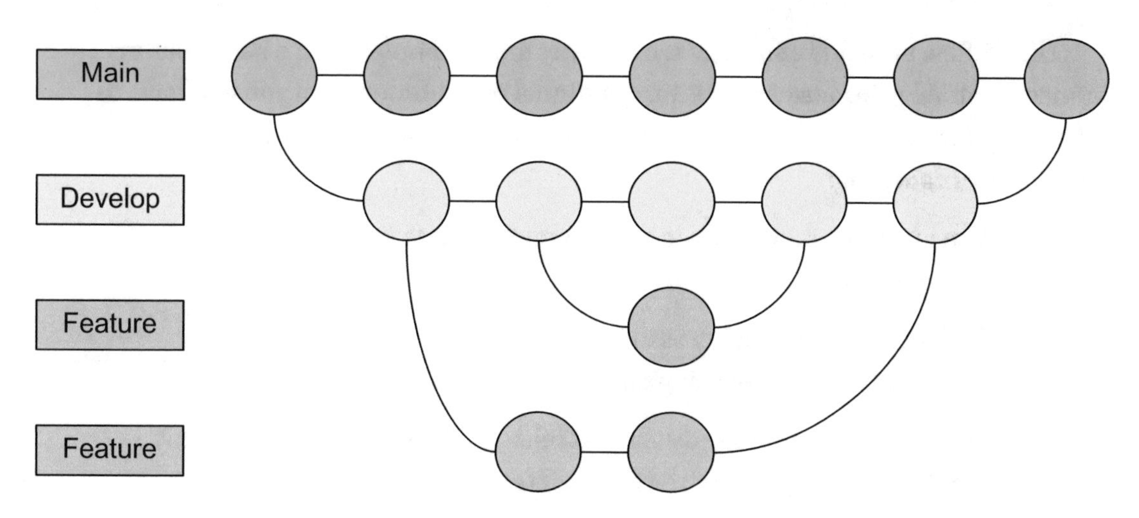

Figure 10-3. *Example of branching with Gitflow*

There are three dbt environments that you will need to have set up when it comes to using Gitflow. Much like the Feature Branch workflow that we discussed earlier, the main branch corresponds with your production dbt environment, and feature branches correspond to your development environment. However, the difference between the two strategies is in the addition of a third environment that corresponds with the develop branch. Frequently, this environment will be named staging, QA, or UAT. In Figure 10-4, we expand on the image of Gitflow branching to show you which environment these branches operate in.

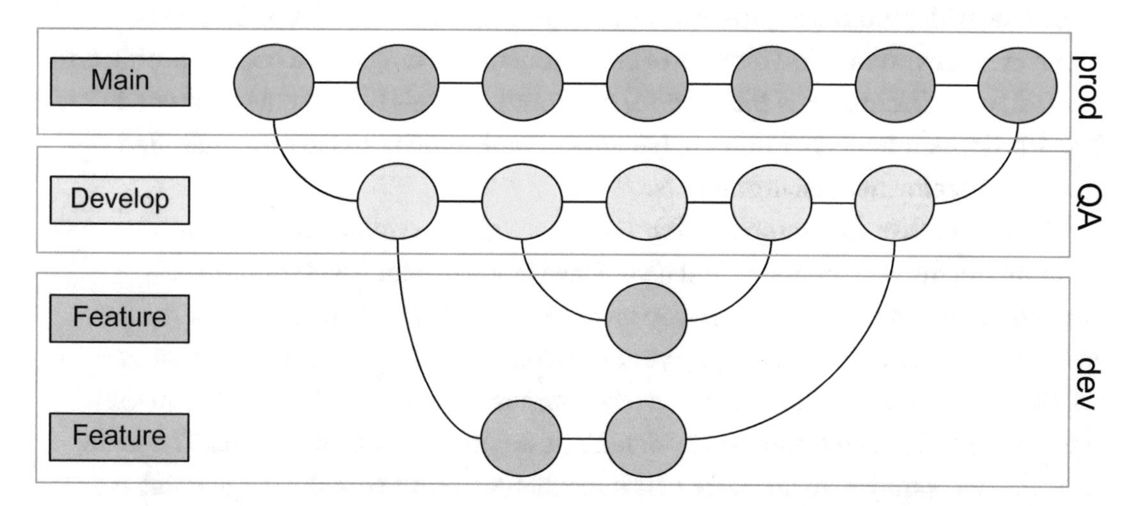

Figure 10-4. *Example of dbt environments when branching with Gitflow*

The Gitflow branching strategy is obviously more complex than a basic feature branching strategy because there is an additional environment that you will have to maintain. Here are some of the reasons why you might want to choose a Gitflow strategy over feature branching:

- You want to control the production release cycle instead of having automated releases.
- You want to have quality assurance and user acceptance testing completed before releasing changes to production.
- You need to thoroughly document releases. This is frequent in highly regulated industries such as healthcare and finance.

Environment Management with dbt Cloud

Now that you understand what environments are when developing with dbt and how environments should correspond with your git workflow strategy, let's cover how environments are created and managed within dbt Cloud. To start, dbt Cloud only has the concept of two types of environments: development environments and deployment environments. In dbt Cloud, you can only have one development environment. This environment typically will point to a target database that is configured as a clone of production. This is where developers can run dbt jobs on their code changes without interfering with production jobs. Any other type of environment in dbt Cloud is a deployment environment. This type of environment can range from being a production, QA, UAT, or CI (Continuous Integration) environment. This type of environment differs from the development environment because organizations can have an unlimited number of deployment environments.

Fortunately, we already showed you how to configure your development environment during Chapter 2, and if you have followed along with examples throughout the book, this is where you have been working! Setting up a deployment environment is as easy as setting up your development environment. In fact, since dbt Cloud only allows you to have one development environment, any subsequent environments that you create will by default be deployment environments. The main thing that we want you to be aware of is that when you configure the target database for this environment, you shouldn't use the same database as your development environment. By having separate databases for each environment, you ensure that you can follow a reliable code promotion process, and code that is not production ready will not run against the production database. Common naming conventions for a production database include

- Prod
- Production
- Analytics (this is the most user-friendly)

In addition to creating your deployment environment, you will also manage your environment variables from the dbt Cloud UI. If you recall from Chapter 6, we talked about how to use the env_var Jinja function, but we also showed you how to set up environment variables. Keep in mind that each environment variable created in dbt Cloud can have a project default to fall back to if you haven't configured the variable for a specific environment.

Environment Management with dbt Core

If you're a user of dbt Core, then the responsibility of environment management is handled differently than dbt Cloud users. dbt Cloud users manage all over their environment configuration and environment variables directly from the web UI, but dbt Core users will need to handle this in a much different way. Let's first look at how to configure different environments when using dbt Core, and then at the end of the section, we will talk about environment variables.

If you recall from Chapter 2, we mentioned that when you run the dbt init command a series of questions will show up in the terminal that enable dbt to generate a *profiles.yml* file for you. This file is where your connection profiles are stored. A connection profile is a set of configurations that tell dbt how to connect to your target database. When you run dbt, it will search your *dbt_project.yml* for the profile name's value and then use that value to look up the correct connection profile in the *profiles.yml* file. We've mentioned this before, but to reiterate when you run dbt, it will search for this file within the current directory of your terminal, and if it can't find it, it will instead look in the *~/.dbt/* directory.

Note: When you run dbt init, the profiles.yml will be created for you and stored in *~/.dbt/profiles.yml.*

A connection profile is made up of one or more **targets**, which are commonly used to provide different connection configurations depending on the environment you're running dbt in. A good way to remember the difference is that a connection profile corresponds to a project, and a target corresponds to an environment. In Figure 10-5, you can see an example of how we have configured two targets, dev and prod, within the my_first_dbt_project profile.

```
1  my_first_dbt_project:
2    target: dev
3    outputs:
4      dev:
5        type: snowflake
6        account: super.secret.account
7        database: dbt_learning
8        user: super_secrect_username
9        password: super_secrect_password
10       role: dbt_transformer
11       schema: public
12       threads: 8
13       warehouse: dbt
14
15     prod:
16       type: snowflake
17       account: super.secret.account
18       database: dbt_learning_production
19       user: super_secrect_username
20       password: super_secrect_password
21       role: dbt_transformer
22       schema: public
23       threads: 8
24       warehouse: dbt
25
```

Figure 10-5. *Example profiles.yml with two targets*

Also, notice on line 2 that we instructed dbt to use the dev target, so by default when we run any dbt commands, they will be executed using our developer credentials, and models, seeds, etc. will be materialized in our dev (dbt_learning) database. This is useful for development; in production, we of course want dbt to execute jobs against the prod target. This can be achieved by passing the desired target name to the target command-line flag with any dbt command. For example, if you wanted to build your entire dbt project as part of a production job, you would structure the command like this: `dbt build --target prod`.

As we discussed in Chapter 6, you can make a few changes to your *profiles.yml* by removing hardcoded secrets and replacing them with the env_var Jinja function. We recommend doing this anytime you are going to deploy dbt because you don't want your secrets checked into source control. However, you can also use the env_var Jinja function anywhere else in your dbt project, but when using dbt Core, how do you get these variables to be available for use by the env_var function? Well, this is a bit of a complicated question because it ultimately depends on *where* you will be deploying dbt. The simplest way to make your environment variables available is to use a *.env* file. This is a common pattern, but not the only one that you can use. You could also use a secret manager, such as AWS Secrets Manager, if you will be deploying within AWS infrastructure. Regardless, you will need to think through how you will make your environment variables available within your production environment.

Production Jobs

Once you have set up a production environment, you're now ready to start deploying jobs to production. When we refer to a dbt job, we are talking about the running of one or more dbt commands in a production environment. Typically, jobs are triggered by a scheduler such as cron, events from upstream processes such as the completion of a data ingestion pipeline, or by an orchestration platform. There are many different ways which a job can be triggered, and these vary based on your deployment of dbt. In this section, we will cover creating jobs and triggering jobs in dbt Cloud and dbt Core.

dbt Cloud

dbt Cloud offers a variety of ways that you can create, monitor, and trigger jobs. In this section, we will focus on how to create jobs using the dbt Cloud web application. To get started, you can create a job directly within the web application from the jobs page by selecting the "create job" button. From this page, you will be presented with a few options including naming the jobs, selecting the environment it will run in, defining the commands that will run, and selecting what will trigger the job.

When you are defining your environment configuration for a job, you have the ability to select which environment your job should run in, which version of dbt this job should run (or you can just inherit the environment's version), and set the target's name. Recall from Chapter 6, we talked about the target variable and how it provides you access to your environment's configurations, which can be useful to alter the behavior of your dbt jobs depending on which environment you are running.

Following this, you can configure the execution settings that allow you to do things like define a timeout so that the job is killed if it takes too long, but most importantly this is where you will list the dbt commands that will be executed by your job. In Figure 10-6, you can see an example where we have listed four commands to be run as part of this job: dbt seed, dbt snapshot, dbt run, and dbt test. Really, this could be simplified to one command, dbt build, but we wanted to demonstrate the ability for you to be able to run an arbitrary amount of dbt commands as part of your job.

Figure 10-6. *Example of configuring a dbt Cloud job with multiple commands*

The final, and arguably most important, part of creating a dbt Cloud job is defining the trigger for the new job. Currently, dbt Cloud offers three categories of triggers for jobs:

- Scheduled
- Continuous Integration
- API

For the rest of this section, we will focus on schedule- and API-triggered jobs, but we will trace back to Continuous Integration jobs later in the chapter when we will discuss CI/CD in greater detail. Starting simple, jobs can be triggered on a schedule from dbt Cloud in two ways. The first way is to use their simple scheduling tool to select which days the jobs should run and the interval it should run on. For example, you could select that the job should run Monday-Friday once per hour. However, if this isn't robust enough to meet your scheduling needs, you can also provide a cron schedule that dbt Cloud will use to trigger your job. If you are new to cron scheduling, or need a refresher, Figure 10-7 provides the cron syntax.

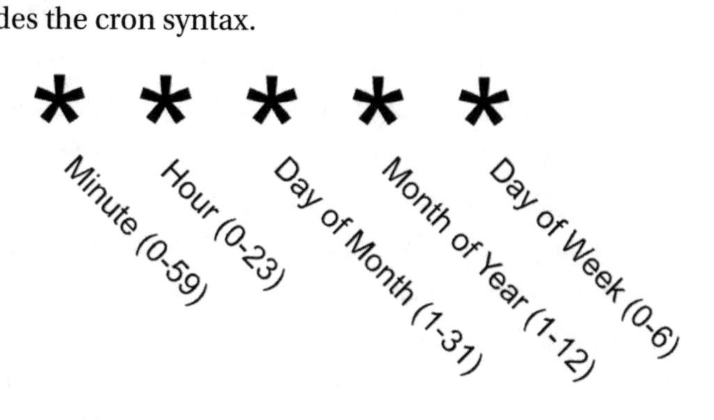

Figure 10-7. *Cron syntax*

The second and most flexible way to trigger dbt Cloud jobs is using the API. Using the dbt Cloud API, you can do things like

- Create, update, or delete a job
- Get the run status of a job
- Trigger a job
- Administrative tasks (list users, update licenses, list projects, etc.)

By using this API, you can trigger your dbt Cloud jobs programmatically. A common approach is to use an orchestration tool, such as Airflow, to hit the endpoints in this API following the completion of upstream tasks in a data pipeline. For example, suppose that you have an Airflow DAG that has three steps:

1. Ingest data from a CRM
2. Ingest data from social media ad platforms
3. Trigger dbt upon the completion of tasks one and two

In this pipeline, task three would make an API call to dbt Cloud to trigger the job that you have defined. In fact, using Airflow to trigger dbt Cloud jobs is very simple because there are already Airflow Operators that abstract the complexity of having to trigger dbt jobs and poll for the status. But, Airflow isn't the only orchestration tool that you can seamlessly integrate dbt Cloud with. You can also very easily integrate with Azure Data Factory, Prefect, and Dagster to trigger your dbt Cloud jobs as a task in a data pipeline that contains tasks beyond dbt jobs.

dbt Core

When it comes to running dbt Core jobs in production, you have more flexibility than you do with dbt Cloud, but that comes with the trade-off of complexity. Running dbt Core in production is more flexible because you can run it pretty much anywhere that you can deploy software, and you can trigger your jobs in any way that you can imagine. However, the trade-off of complexity arises because you are not only responsible for running your jobs, but you also have to be concerned with the infrastructure that you use to run your jobs. When you use dbt Cloud, the concern of infrastructure has been abstracted away as it should be for a paid SaaS solution.

While setting up infrastructure to run dbt Core might sound intimidating at first, you can actually make it quite simple! And for many teams, keeping things simple is a great place to start. Unless you have unlimited resources on your team, we recommend that you start with a simple deployment strategy of dbt Core and progressively make it more complex as the need for additional flexibility arises. Table 10-1 provides some common deployment patterns for dbt Core and the flexibility and complexity that is associated with them relative to other strategies.

Table 10-1. *dbt Core deployment complexity vs. flexibility*

Strategy	Complexity	Flexibility
GitHub Actions	Low	Low
Virtual Machine	Medium	Low
Containerized	Medium	High
Orchestration Tool	High	High

Each of these strategies gets progressively more complex, but there do come times when it makes sense to move forward with a more complex deployment strategy. If we were building a data team from the ground up and implementing dbt Core, we would keep things simple and start with using GitHub Actions to run our production jobs. This strategy is extremely easy to deploy because you don't have to worry about managing any infrastructure. GitHub Actions are run inside a virtual machine that GitHub will spin up on your behalf when your workflows are triggered. Additionally, GitHub Actions can be triggered using a cron schedule, and if this interests you, we've included an example workflow in the book's Git repo.

If GitHub Actions aren't an option for you, then you could run dbt by directly installing it in a virtual machine and creating a cron schedule on that machine. However, we don't recommend this strategy because you will have to manage the virtual machine. Additionally, dbt is not resource intensive at all because it is simply compiling your SQL and Python code to be shipped off to your data platform, so as a result virtual machines tend to be overkill for running dbt in production. Because of the low flexibility that a virtual machine offers, we recommend that you jump up the ladder in complexity and use a containerized deployment and/or an orchestration platform.

The final two strategies for running dbt Core jobs in production, containerized deployment and orchestration platforms, aren't actually mutually exclusive. If you're familiar with running software in containers, then you'll know that you need something to orchestrate those containers, such as Kubernetes or AWS Elastic Container Service (ECS). In addition to this, you may include a data orchestration platform to trigger a containerized deployment of dbt Core. If you are interested in learning more about containerized deployments, then a great resource is Shimon Ifrah's book (2019) *Deploy Containers on AWS*.

An example of this would be to have dbt Core inside of a Docker container, have that container included in an AWS ECS task, and have Airflow trigger the ECS task. We really would only recommend deploying dbt Core within a container and/or triggered by a data orchestration platform when you have external factors that you want to use to trigger dbt, or when you want downstream tasks to be triggered when dbt runs are completed. For example, you need your data orchestration platform to trigger a data refresh in your Business Intelligence tool once a dbt run completes. If you have this use case, or something similar, then the additional complexity of this deployment strategy may make sense.

As you can imagine, you can quickly make your deployment of dbt overly complex, so we encourage you to think through the trade-offs of your strategy when designing it. Things to keep in mind as you design your deployment strategy should include

- Complexity vs. flexibility.
- Remembering that dbt is not resource intensive, so you don't need to run it on a 512GB Bare Metal machine. A micro EC2 instance is more than sufficient in most cases.
- Should dbt be triggered by external upstream tasks?
- Should dbt trigger external downstream tasks?

- Does your team have the bandwidth to support a complex deployment?
- Does a more complex deployment strategy add additional value to your business?

CI/CD

One of the core philosophies of dbt is to bring software engineering best practices to the world of data transformations. Throughout the book, we have talked about some of these best practices including collaborative development, testing, version control, and documentation. To extend this idea, we want to spend this section focusing on CI/CD, which is arguably one of the most important best practices in software development. CI/CD, or Continuous Integration/Continuous Deployment, is the process of automating the testing and deployment of software, but the same ideas can be extended to dbt projects. The goal of CI/CD is to ensure that changes to your codebase are automatically and reliably integrated into production so that teams are able to deliver new releases of their code faster.

CI/CD is broken up into two fundamentally different components, with the first being Continuous Integration which aims to catch bugs, merge conflicts, or other integration issues before developers merge their changes into the stable production codebase. This is typically achieved by running builds and tests against the changed code. Each CI pipeline can be triggered at different times, but one of the most common is to trigger the pipeline when a pull request (PR) is opened by a developer. Within the context of developing dbt projects, a CI pipeline will typically invoke dbt to build and test changes to models, seeds, and snapshots whenever a PR has been opened. We will discuss strategies for designing CI pipelines for dbt later in this section.

The second part of CI/CD is Continuous Deployment, where the release of code changes and deployment of those changes have been automated. Continuous Deployment typically happens after Continuous Integration pipelines have passed and a developer has merged their changes into the production codebase. CD enables frequent releases because they have been automated, so the need to wait for formal releases has been eliminated when CD is implemented. To implement Continuous Deployment successfully, teams typically rely on various tools such as version control systems such as Git, build tools (e.g., Jenkins and CircleCI), infrastructure-as-code such as Terraform, and deployment automation scripts. It's essential to have robust testing strategies, including data quality tests and unit tests, to ensure the reliability of the deployed code changes.

Overall, CI/CD can be thought of as a process to automate code changes from development to production environments by building and testing changes, automated releases and deployments, and monitoring of deployments to ensure that they are stable. For dbt codebases, we've found that implementing CI/CD is a desirable practice to follow. For the remainder of this section, we will focus on building different components of CI/CD pipelines in both dbt Cloud and dbt Core.

Setting Up an Environment for Continuous Integration

There are two schools of thought when it comes to running CI pipelines and environments; the first is to just use your production environment, and the second is to set up a dedicated CI environment. While you can, and many people do, use their production environment to run their CI pipelines, we recommend that you create a dedicated environment to run your dbt CI pipelines. In our opinion, this is the most ideal setup because it best separates the concerns of production workflows and Continuous Integration workflows.

Setting up a Continuous Integration pipeline in dbt Cloud is very simple because you just need to create an additional deployment environment. In the past, we have configured CI environments to use a different set of credentials than our production environment. This has proved to be better for monitoring because we could easily track down which user executed queries against the database.

When it comes to setting up an additional dbt Core environment for CI, you will need to alter the *profiles.yml* file and add an additional target to your project's profile. Figure 10-8 shows an example of how you might configure this target. There are additional changes that you might want to make to your CI environment depending on how you want your CI pipelines to run. We will discuss the changes you may want to make shortly.

```
1  my_first_dbt_project:
2    target: dev
3    outputs:
4      ci:
5        type: snowflake
6        account: super.secret.account
7        database: dbt_learning
8        user: my_ci_user
9        password: super_secrect_password
10       role: dbt_transformer
11       schema: public
12       threads: 8
13       warehouse: dbt
14
15     dev:
16       .......
```

Figure 10-8. *Example of a CI target in a profiles.yml file*

Running a CI Job Using State Comparison

Imagine that you've just completed building a feature and you want to merge your feature branch into the main branch. Your team has set up a CI pipeline that invokes a full dbt build whenever a new pull request is created. As you can imagine, the run time of this CI pipeline grows linearly with growth of your codebase. This can lead to decreased developer efficiency because you have to sit around and wait for a full build of your dbt project to complete, and pass, every time you create a pull request. Fortunately, dbt offers a node selector called **state** which serves to improve the speed of this feedback loop and conform better to CI/CD practices.

The state method is a node selector that you can use to modify a dbt command, such as **dbt build**, **dbt run**, or **dbt test**. This method allows you to build just the changes that you've made instead of building your entire project. The way that this node selector works is by comparing the *manifest.json* of your changes to the *manifest.json* of a previous version of your project. Within the context of a production CI pipeline, you will be comparing to the manifest from your current stable production deployment. Throughout the remainder of this section, we will look at using state comparison to build CI pipelines that will trigger when a pull request is opened against the main branch. However, a really great read from dbt Labs can be found in this article: `https://discourse.getdbt.com/t/how-we-sped-up-our-ci-runs-by-10x-using-slim-ci/2603`. As with most things related to production deployments, there are vast differences in how CI pipelines are built when it comes to dbt Cloud and dbt Core. Let's first look at how to build this type of workflow in dbt Cloud and then follow this with building a similar workflow in dbt Core. By doing this, we can understand the differences and trade-offs of using the SaaS offering and building your own CI pipelines for dbt.

Note The state method is only one of many node selectors. If you are interested in running more complex and specific dbt jobs, we encourage you to check out Chapter 4's section on "Node Selection."

---

dbt Cloud

dbt Cloud has a built-in feature called Slim CI, which streamlines the process to achieve a more simple CI pipeline for running and testing dbt transformations. Slim CI jobs are configured to trigger when a pull request is opened against your default branch (e.g., main), compare your changes to a deployment job, and are run using the **state** node selector that we just discussed. By setting up a Slim CI job in dbt Cloud, you will only run code changes and their downstream dependencies as part of your CI pipeline.

To set up a CI job using dbt Cloud, you will need to do a few things:

1. Optionally, create a new environment for CI workflows.
2. Ensure dbt Cloud has been connected to your Git platform (e.g., GitHub, Azure DevOps, etc.).
3. Create a new dbt Cloud job.

We've already discussed how you can create an isolated CI environment, which is an optional step anyway, so we will skip forward to making sure that your Git platform has been connected with dbt. Also, when you set up your dbt Cloud account, you should have connected to your Git provider. If you did this and you're using GitHub, you can check the GitHub Apps section of your repository settings, and you should see the dbt Cloud app similar to Figure 10-9. Other Git platforms should have a similar setting to allow you to validate that dbt Cloud is integrated. If you don't see anything showing that dbt Cloud is connected, you should refer back to Chapter 2 to make sure that you have correctly connected dbt Cloud to your Git platform.

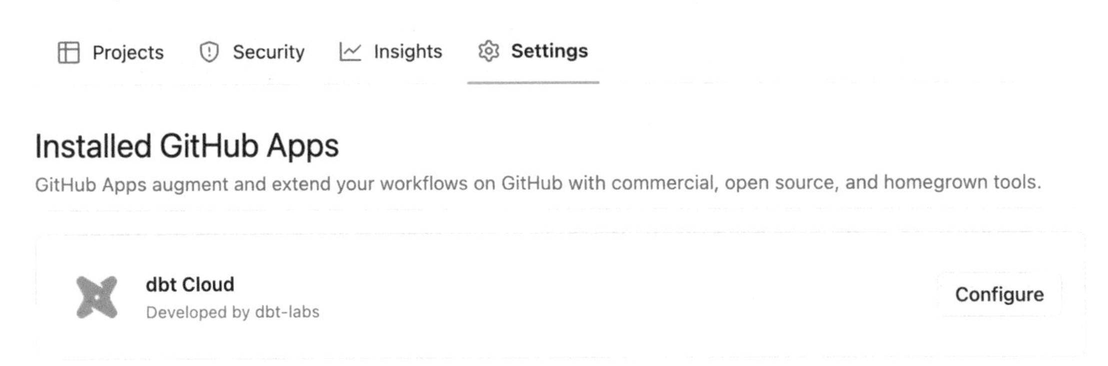

Figure 10-9. *Confirmation that the dbt Cloud is connected to GitHub*

Now that you have confirmed that dbt Cloud is integrated with your Git platform, you should move on to setting up a new job. There are three things in particular that you will need to do that are special about setting up CI jobs in dbt Cloud. The first is to select the job that you want to defer to for state comparison. Within the dbt Cloud UI, there will be a drop-down menu, named *Defer to a previous run state?*, that allows you to select another job from your account. For the purposes of CI, you should select the job of the environment you are merging your changes into. So if you are merging from a feature branch to main, then you should defer to your production job. When this CI job is triggered, dbt Cloud will fetch the manifest of the job you select to defer to so that it knows which models, seeds, etc. it should build as part of the run.

Once you have selected which job to defer to for state comparison, you will need to define the command that you want dbt to run as part of your CI job. For the sake of simplicity, we will just use the **dbt build** command, but as you might have guessed, we need to alter this command slightly so that it only builds changes. To do this, we can alter the command to be

```
dbt build --select state:modified+
```

This command starts off with a typical **dbt build**, but it is then modified by using node selection where we've used the **state** node selector. We have instructed dbt to only build modified nodes and their downstream dependencies (indicated by the plus sign). As you can imagine, it's useful to build downstream dependencies so that you are confident that your changes don't break anything unintentionally.

The final step is to set the trigger for this job. If you recall from earlier in the chapter, there are three triggers that can be used to kick off dbt Cloud jobs, and the one that we haven't discussed yet is the Continuous Integration trigger. To use this trigger, you will select the box name *Run on Pull Request*, and as the name implies, this job will now trigger whenever a pull request is opened. Behind the scenes, dbt Cloud is listening for webhooks coming in from your Git platform that provide events related to pull requests being opened, updated, closed, etc. Whenever dbt Cloud receives one of these webhooks, it will trigger your CI job. If you use GitHub, you will see a status check show up in your pull request from the dbt Cloud GithHb app showing the status of your CI job, which can be seen in Figure 10-10.

***Figure 10-10.** Example of dbt Cloud CI job status on a GitHub PR*

Once you have completed all of these steps, you have successfully set up your first dbt Cloud CI job! Using Slim CI is a great way to improve the run times of your CI jobs and keep changes out of production until you are confident that they are stable. Additionally, there are recent updates to the dbt Cloud Slim CI process that further improves optimization by autocanceling jobs that have become stale. This could help save on your cloud data warehouse bill because dbt won't be utilizing unnecessary resources. Before we move on, we want to talk about where dbt build models in your database for Slim CI jobs. You might have already been wondering this because we didn't make any mentions of setting up an additional database or schema for dbt to build in, and this is because dbt Cloud will actually handle this for you as part of the Slim CI process. Whenever a new CI job is triggered in dbt Cloud, a new schema in your environment's target database will be created. The schema will follow the naming convention, **dbt_cloud_pr_<job id>_<pr id>**. An attempt will be made to build all of your changes within this schema, and then when the job has completed, dbt will attempt to drop this schema.

At the time of this writing, there are some known limitations related to the temporary schema that dbt Cloud generates, including

- The schema not dropping after a job completes
- Some changes being built in your target schema(s) instead of the PR schema

Both of these limitations typically relate back to you having changed the **generate_schema_name** or **generate_database_name** macros. To solve this problem, you can alter these macros further to account for the temporary PR schema with logic similar to this (may need to alter depending on your custom schema logic):

```
{%- set default_schema = target.schema -%}
{%- if custom_schema_name is none -%}
  {{ default_schema }}
{%- elif target.name == 'ci'-%}
  {{ default_schema }}_{{ custom_schema_name | trim }}
{%- else -%}
  {{ custom_schema_name | trim }}
{%- endif -%}
```

dbt Core and GitHub Actions

If you're running dbt Core, then unfortunately you won't have access to run your CI jobs using the dbt Cloud app that integrates directly with different Git platforms such as GitHub. Instead, you will need to recreate this process using a CI tool. When it comes to creating a CI job for dbt Core, we will take many of the same approaches that dbt Cloud does. For the example in this section, we will create a GitHub Actions workflow file that achieves most of the same functionality that you get with Slim CI in dbt Cloud.

The biggest difference between the process we will show you and the way the dbt Cloud implements CI jobs is that instead of creating a temporary PR schema, we will create a clone of our entire target database, run the CI job in this database, and then drop the database after the CI job completes. Unfortunately, not all cloud data platforms support cloning entire databases, but Snowflake does. If you are using a different data platform, you will need to think through this process a bit more as you may need to alter some of it to produce temporary schemas instead. However, if you are using Snowflake, we prefer this approach because it keeps the CI jobs and production jobs an extra step away from each other since they run in separate databases. Furthermore, this approach is our preferred method because it is much more simple to implement than generating temporary schemas. Not to say that generating temporary schemas is difficult, but we are all about efficiency and time to value.

At a high level, we will need to build three new files and make a change to one existing file:

1. Create a macro to generate the PR database.
2. Create a macro to drop the PR database.
3. Update the *profiles.yml* to point to the CI database.
4. Create a GitHub Actions workflow file.

For the first two steps, we will need to create a couple of macros. Listings 10-1 and 10-2 provide the code for the macros to generate and drop the PR database using Snowflake's cloning feature. Notice in both of these macros we reference the target.database and append a value from an environment variable name DBT_PR_NUMBER to the end of the database name. When we build the GitHub Actions workflow, we will point these macros to the production target and pass in a variable with the PR number. This will effectively create a clone of our production database with one database per unique pull request.

Listing 10-1. A macro that clones your target database for CI

```
{% macro clone_database_for_ci() %}
  {% set sql='create or replace database ' + target.database + '_' + env_
var('DBT_PR_NUMBER') + ' clone ' + target.database %}
  {% do run_query(sql) %}
{% endmacro %}
```

Listing 10-2. A macro that drops your temporary CI database

```
{% macro drop_database_for_ci() %}
  {% set sql='drop database ' + target.database + '_' + env_var('DBT_PR_
NUMBER') %}
  {% do run_query(sql) %}
{% endmacro %}
```

Before we move on to creating the workflow file, let's update the *profiles.yml* to point to the database that we've cloned instead of pointing to the production database. To do this, we will update the database name to be appended with the DBT_PR_NUMBER environment variable, as seen in Figure 10-11. This will dynamically update the CI target with the correct database for each invocation of the CI pipeline that we are building.

```
26  ci:
27    account: YOUR_ACCOUNT_HERE
28    database: analytics_{{ env_var('DBT_PR_NUMBER')}}
29    password: YOUR_PASSWORD_HERE
30    role: dbt_transformer
31    schema: public
32    threads: 8
33    type: snowflake
34    user: YOUR_USER_HERE
35    warehouse: dbt
```

Figure 10-11. *Updated CI target pointing to the CI database*

Lastly, we need to create the workflow file that GitHub will use to run our CI pipeline. GitHub looks for a directory called *.github/workflows* and will generate workflows based on the YAML files within it. To make our CI pipeline work, there are several steps that we need to include in our workflow file:

1. Check out the main branch.
2. Set up Python in a virtual machine.
3. Install dependencies for our project.
4. Run dbt deps against the main branch.
5. Run dbt compile against the main branch.
6. Copy the production target folder to be used for state comparison.
7. Check out the PR branch.
8. Run dbt deps.
9. Run the `clone_database_for_ci` macro.
10. Run dbt build using the state node selector.
11. Run the `drop_database_for_ci` macro at the end.

There are a lot of steps here, but we mainly can focus on steps 9–11. Steps 9 and 11 use the dbt run-operation command to execute the `clone_database_for_ci` and `drop_database_for_ci` macros, respectively, so that the temporary CI database is cloned for us. We will point the command in step 9 to the production target so that it uses the correct target.database when creating the clone.

Note If you refer to the workflow file, you will see that we have made DBT_PR_NUMBER available as an environment variable. dbt will automatically set this environment variable, but you can pass others in via GitHub by storing secrets: `https://docs.github.com/en/actions/security-guides/encrypted-secrets`.

Next, we will run dbt build using node selection, but this command will be a bit more complicated than the one we ran in dbt Cloud. It is slightly more complicated because we have to tell dbt where we've stored the state files that we want to compare to, and we need to tell dbt which target to use. Listing 10-3 shows an example of what this command will look like inside of the workflow file.

Listing 10-3. dbt run using node selection inside the workflow file

```
dbt build \
--select state:modified+ \
--state ./production_target/target \
--profiles-dir . \
--target ci
```

If you are interested in running this workflow in your project, you can find the full workflow file in this book's repo at *~/.github/workflows/build_dbt_diffs_ci.yml*. As you can see, running a CI job with dbt Core is more complex than it is with dbt Cloud. Building your own CI pipeline instead of using dbt Cloud requires more initial setup, but typically once you have surpassed the setup curve, then these workflows tend to just work. If you aren't comfortable with building GitHub Actions, or another Git platform's equivalent, then we encourage you to consider using dbt Cloud.

Linting

Many people have their own preferences when it comes to formatting code, naming conventions, or other coding standards. In our experience, people who write SQL have some of the most opinionated preferences when it comes to how to format code. Some examples include

- Trailing vs. leading commas
- CAPITAL vs. lower keywords
- CamelCase vs. snake_case

The list goes on, but there is no sense in wasting valuable developer time when it comes to these stylistic differences. Instead, your organization should have a standard style guide for the codebase to set the precedence on how all code committed to the main branch should be styled. In the world of software development, linting refers to the process of analyzing code for stylistic issues, adherence to coding standards, and inconsistencies across the codebase. A linting tool, also referred to as a linter, operates on a set of predefined rules to lay out the coding standards and best practices for your repository. By incorporating linting into the development process, you and your team can improve code quality, enforce coding standards, catch errors early, enhance collaboration, and promote maintainability of the codebase. Linters help establish consistency and reliability for stylistic decisions in the codebase.

We strongly believe that you should implement a linter as part of your Continuous Integration pipeline because it helps maintain a high standard for the stylistic quality of code that enters your production codebase. One of the most frequently used SQL linters is called SQLFluff. SQLFluff is extremely customizable and can help you solve for all of the stylistic differences we mentioned earlier, and more! SQLFluff checks for adherences to what they call *rules*, and here is a list of some of those rules:

- Enforcing table aliases
- Preventing ambiguity (e.g., require inner join instead of join)
- Enforcing a keyword capitalization standard
- Requiring coalesce instead of ifnull
- Preventing the use of subqueries and requiring CTEs instead

The list goes on and on, but we do encourage you to implement SQLFluff. We also like to have SQLFluff triggers as part of our CI pipelines so that we can be sure that the proper linting rules are adhered to before a pull request is merged into the main branch. We've included a GitHub Actions workflow file in the book's repository that you can use to trigger SQLFluff linting when a pull request is opened. Additionally, there are other open source tools that you could work into your CI workflows such as dbt-checkpoint and dbt-project-evaluator. We've discussed these earlier in the book, but we do think it is worth considering adding these to your automated checks.

Continuous Deployment

When it comes to Continuous Deployment and dbt Cloud, you don't even have to think about how your production-ready code becomes available to your jobs that you've created. Instead, dbt will simply fetch the most recent state of your production codebase when each job begins in dbt Cloud. This streamlines the deployment process of dbt because it requires no effort on your part with regard to Continuous Deployment.

This is much different when it comes to running dbt Core deployments, because you haven't paid a SaaS company to handle the deployment for you. Instead, it's up to you to build a Continuous Deployment process so that when new releases of production code happen, those changes are available in your dbt jobs. This process can vary vastly depending on how you have deployed dbt Core. For example, if you've deployed dbt Core using GitHub Actions, then this process is quite simple because you will simply check out your production codebase each time the GitHub Action is triggered. So, not much to consider with this deployment approach, but if you have a more complex approach such as a containerized deployment, you will need to have an automated pipeline that completes steps such as

- Building new containers and pushing them to a container repository
- Making them available to your container orchestration tool (K8s, ECS, etc.)
- Redeploying your data orchestration tool with the updated dbt code available to it

Because of how many different deployment options there are with dbt Core, we can't cover the steps to build Continuous Deployment pipeline for each, but we do want you to be aware that it is something you need to consider when thinking about running dbt Core in production.

Final Deployment Considerations

Throughout this chapter, we have covered many considerations that you need to think through when deciding whether you want to run your production deployment in dbt Cloud or design your own deployment using dbt Core. We've covered some of the key components such as environment management, triggering jobs, and CI/CD pipelines, but there are additional considerations that are easily overlooked if you are new to managing a production deployment of dbt. In this final section, we will cover deploying dbt documentation, accessing logs, and alerting/monitoring.

Documentation Deployment

In Chapter 9, we covered how generated documentation works in dbt and primarily focused on how you can easily deploy documentation to dbt Cloud. Within dbt Cloud, it is as simple as toggling a button on in one of your production jobs to instruct dbt Cloud to generate and publish new documentation for you. If you use dbt Core, this process becomes more complex. We covered a shortlist of ways that you can deploy your documentation when using dbt Core, but we didn't go into much detail. Two simple approaches to deploying your dbt documentation include publishing as a static site on AWS S3 or publishing it to Netlify.

Both of these options are common approaches and range a bit in complexity. First, deploying to Netlify is the simplest approach because you can just push to documentation here via their command-line tool, and if you pay for Netlify, you can set up authentication. This puts your documentation behind a wall of security without much administration overhead requirement on your part. The second option, deploying to AWS S3, is also very simple because you can deploy using the AWS CLI, but this comes with the trade-off of more complexity when it comes to putting your documentation site behind some sort of authentication. However, you need to decide the best method for securing your documentation site if you choose to host it on AWS S3.

A final consideration for deploying documentation on your own instead of letting dbt Cloud handle it is the ability to customize your documentation. If you deploy the documentation yourself, you can override pretty much anything you want to. In our experience, we have enjoyed doing this because you can update the CSS to put your company's logo at the top of the documentation instead of having the dbt logo, and you can override the colors of the documentation to make them more on brand with your company. This of course adds complexity, but if it is something that interests you, it is definitely something that can be done. GitLab has open sourced their dbt project, and we would recommend that you check out this, `https://gitlab.com/gitlab-data/analytics/-/tree/master/transform/snowflake-dbt/docs`, directory in their repository to see how they have customized their dbt documentation.

Monitoring and Alerting

Imagine that you have a production dbt job that is supposed to run once an hour on the 15-minute mark to update a business-critical dashboard. But, it's now 1:30 pm, and you get a message from one of your stakeholders letting you know that the dashboard hasn't been updated this hour. A stakeholder reaching out to you to let you know that data has become stale or that data is inaccurate is not an ideal method for monitoring your data transformation pipelines. It would be much better to think through this before deploying business-critical data products to production. Alerting and monitoring is one of the most important considerations that you need to think through when it comes to a production deployment of dbt.

There are many categories of alerting and monitoring, but let's focus on three:

- Data freshness
- Data quality
- Job results

The first piece of monitoring that you should implement is data freshness checks, which you can configure dbt to validate against your sources. This is useful because it can help you identify the root in the scenario like we have where a stakeholder has mentioned a dashboard is out of data. With source data freshness checks, which we discussed in Chapter 3, dbt can throw an error or a warning when your source data is out of date. Without these checks, you could have silent failures in your dbt runs because the transformation jobs will run as normal, but they won't be processing any data.

Next, you should have data quality checks implemented. We talked about data quality checks at length in Chapter 8, but we want to reiterate the importance of having these implemented prior to going to production. Data quality tests also help to prevent silent failures by making sure that your assumptions about the expected output of the data are valid. These tests can range from not null tests, uniqueness tests, and more complex tests such as monitoring from unexpected trends in the data.

Lastly, you should be monitoring your job results because these are where you will catch the loud failures. In dbt Cloud, you can configure emails and slack notifications to be triggered based on certain criteria for jobs. Suppose that you want to alert a slack channel each time a production job fails. This is useful information to have, but it's even better when you can get additional context about failures. This is where monitoring tools come in to help out. For example, you might want to configure your dbt artifacts to be sent to Datadog, Metaplane, or something similar to send alerts with context when failures happen. Tools like this can also make sure that you are alerted if your jobs are running late, don't run at all, or are taking a longer time to run than normal. Additionally, you might consider setting up a tool like PagerDuty so that your on-call team members can be alerted if jobs are continuing to fail. However, if you choose to send alerts to PagerDuty, we encourage you to be very selective about what triggers an on-call incident. If you aren't, it is very easy for alerts to become noisy and eventually ignored all together by your team.

Summary

Throughout this book, we've taken you all the way from square one of setting up dbt in a development environment and learning many of the major components of dbt to ending with this chapter where we have talked about the considerations you make when deploying dbt to production. We've placed heavy focus on comparing dbt Cloud to designing your own dbt Core deployment strategy because there are some heavy trade-offs when you take one approach over the other. If we were to summarize the differences in one word, it would be: simplicity. If you choose to deploy to production using dbt Cloud, then you defer the burden of deployment complexities to dbt, but this of course comes at a cost because dbt Cloud is a paid SaaS product.

When it comes to deploying dbt Core to production, your options for deployment are much more broad and flexible. But, in contrast to dbt Cloud, this comes with the trade-off of complexity. However, we don't think that you should shy away from a dbt Core deployment if

- Your team is small.
- You have savvy DevOps people on your team.
- You want full control over your deployment.

Regardless of whether or not you choose to deploy with dbt Cloud or DIY-ing it, you will need to understand how to set up production-grade jobs. You'll need to consider the frequency at which your jobs need to run and whether they will be scheduled or will be event triggered.

Before you deploy to production, you should also consider implementing CI/CD practices so that you can be confident in the quality of your code before deployment. Common CI/CD pipelines include running and testing dbt code that has been added or changed as part of a pull request, and many teams choose to implement linting as well. You can make CI/CD as complex or as simple as you desire to, but we caution you to only increase complexity when there is an increased benefit.

Last, but certainly not least, you will want to make sure you have some level of monitoring and alerting set up for your project. If you don't and jobs fail, and they will fail, you pose the risk of ruining the trust in your data products. So, it is always better to be ahead of these inevitable failures and build in alerting so that you know of failures before your stakeholders do!

Index

A

B

C

D

C. Cyr and D. Dorsey, *Unlocking dbt*, https://doi.org/10.1007/978-1-4842-9703-2

E

F

G

H

I

J

K

L

M

N

O

P, Q

R

S

T

U

V, W, X

Y, Z

Printed in the United States
by Baker & Taylor Publisher Services